THE BUILDING CONSERVATION DIRECTORY
SPECIAL REPORT ON HISTORIC CHURCHES

Eighth Edition 2001

ISBN 1 900915 18 9

**PUBLISHED BY**
Cathedral Communications Limited
High Street, Tisbury, Wiltshire SP3 6HA
Tel 01747 871717 Fax 01747 871718
E-mail bcd@cathcomm.demon.co.uk
www.buildingconservation.com

**MANAGING DIRECTOR**
Gordon Sorensen

**EXECUTIVE EDITOR**
Jonathan Taylor

**PUBLIC RELATIONS DIRECTOR**
Elizabeth Coyle-Camp

**ADVERTISING SALES**
Charlotte Dean
Anthony Male
Nicholas Rainsford

**PRODUCTION AND ADMINISTRATION**
Jane Martin
Hannah Moffatt

**TYPESETTING & DESIGN**
Xendo, London

**PRINTING**
Optichrome, Woking



Typefaces: Apolline & Parisine

# contents

FRONT COVER **Caledonia Road church, Glasgow** (Alexander 'Greek' Thompson 1857) at night, brilliantly illuminated for the Crown Street Regeneration Project (photograph ©Anthony Ó'Doibhailein)

# WHOSE CHURCH

Photograph©Geoff Taylor

***St Elisabeth's, Reddish, Manchester**, widely recognised as Alfred Waterhouse's finest church*

## Nigel Hawley

A BUILDING which is important as both a place of worship and as an historic building may be subjected to conflicting requirements, and its maintenance is a heavy burden. Why should the religious community shoulder so much responsibility for the wider community?

Set in a cold draughty historic church, the scene is a side chapel converted by the Victorians into a semblance of vestry accommodation. Here are safes containing registers of Baptisms, Marriages and Burials and the church's family silver. An assortment of wardrobes hold robes for a choir and vestments for a priest, and supplies for Holy Communion. The 20th century contributed a photocopier, now used to propagate *Common Worship* in leaflet form.

By the stained wash-basin (with the only mains water supply in the church), a churchwarden struggles to make tea. It is an interval of a meeting of the church council. Everyone, including the new rector, sits on pink tubular chairs rescued from a closed cottage hospital. Her predecessor had used a grander wooden chair behind a table.

The meeting's first fragment saw the usual business of apologies and minutes pass with unusual speed. There has, for weeks, been a circulation of rumours that change is afoot. Now the rector produces sketches on acetate sheets projected onto plaster, damp from the steaming kettle and bad ventilation.

The question of toilets and kitchen is tackled first, and unsurprisingly gets unanimous approval. The movement of altars is not well received, though all agree that the fine Edwardian screen by Harold Gibbons seriously restricts sight. Issues of disabled access are tackled – the law, it is said, will soon require this. There is also the problem of pews from the 1840s with doors and straight backs. It is suggested that they are good for confining toddlers, but this advantage pales into insignificance beside their amazing lack of comfort, and there is enthusiasm for the purchase of chairs.

# IS IT ANYWAY?...

Photograph©Len Grant

**St Francis, Gorton, Greater Manchester,** Edward Welby Pugin (1866-72).
*Once abandoned, churches are particularly vulnerable to vandalism and natural decay. This magnificent High Victorian Franciscan church and monastery was vacated in 1989 as its dwindling congregation could no longer afford its upkeep.*

Photograph©World Monuments Fund

**Detail of a vandalised altar at St Francis.**
*The monastery and its church were added to the list of 100 most endangered monuments by The World Monuments Fund in 1998, and a trust has now been established to secure its future with the help of a grant from the Heritage Lottery Fund.*

In contrast, there is no enthusiasm for rescuing the endangered 14th century wall-paintings in the south chapel. A conscientious churchwarden has obtained estimates for their restoration, but the church council's present unusual liquidity – financed by a one-off sale of land – would be exhausted by such a conservation project. Voices shout for grants. The Vicar replies that the only grant available will be a contribution from English Heritage to make sound the chapel roof.

And so the moral reckoning begins. Is the Church here to conserve the past or serve the present community? The parish is small, the population thin. The ravages of foot and mouth have left many on the bread line. Generally, and without malice, a feeling emerges that to exhaust funds on the conservation project would mean that the church would lack facilities for the foreseeable future. It may have to close.

The new rector is a person of considerable skill. In the weeks ahead, her acetate sheets become provisional plans for radical reordering. A scheme emerges which constitutes, within the limits of the old arcades and volume, a marvellous liturgical space with a proper provision for human needs and social functions.

The parish architect is, at last, consulted. But it is perceived by the church council that the architect's role is simply one of realising their plans. The arguments have been won, sometimes after painful disagreement and not a few resignations: now is the time for action.

At almost the 11th hour the diocesan advisory committee (the DAC) enters the scene. By now, a feeling of 'ownership' of the project has grown among all but a few members of the congregation. The DAC is seen at best as a legally necessary nuisance; at worst as a body of people totally out of touch with the needs of a struggling parish.

This story could be re-written and made to apply to anything from the 14th century Wool Churches of East Anglia to the 19th century Cotton Churches of the North West.

Where the story goes from here – and I stress, it is not an unlikely story – will depend on the attitudes and diplomatic skills of the DAC, the architect, and the various amenity groups like the *Victorian Society* and the *Ancient Monuments Society* which become involved.

There is often, in situations like these, a clash of interests between those who see the parish church as an historic building to be conserved, and those who have to use it and maintain it as the prime functional resource for a dynamic organism called the Christian Church.

A third party – and one of equal importance in the debate – is the people of the wider local community who do not much attend the church, but for whom the church, in a very real sense, exists. It is theirs and they have no other. It is theirs if they use it or not, or support it or not. Perceived ownership and belonging are key issues in a post-modern society.

It may be difficult for those who are not regular worshippers – and this may include architects, members of DACs and amenity groups, as well as ordinary local people – to feel empathy with calls for reordering. They may perceive the church building as a shrine to history, archaeology, or family memories. Some will look at the building and think 'It has always been like this'.

Yet few of our church buildings are 'all of a piece'. Even the churches of the Tractarians, the English high church reformers of the 19th century, have been enriched by rather more candles, tabernacles and statues than the law would have allowed when they were first built. The same movement transformed all but a few older churches, particularly by the introduction of organs and choir stalls, by the removal of box pews, central pulpits and the other manifestations of the auditory plans of the 17th and 18th centuries.

The reality is that part of the interest and, indeed, the genius of the parish church is that is has been adapted to the Christian Community's changing understanding of liturgy and life down the generations. That sometimes this has involved destruction of the artefacts of previous ages is, in one sense, to be regretted. But I wonder if anyone would genuinely want to sacrifice the layered variety of architecture and monument, furniture and glass which gives so many churches their sense of 'thisness', in favour of a clinical reconstruction of what the church was like – or, more likely, what we today *think* the church was like when the first stones were laid?

There is a further problem about the way the Church (as Christian Community) perceives itself. Christianity grew from

**City landmarks:** *the church spire of St Michael (GP Manners, 1835-37) and the Norman tower of the Abbey tower over the streets of Bath*

Judaism, which, at its heart, had a temple shrine. But central to Christian teaching is that the temple in now no longer something made with human hands, but rather the Christian community itself. "Surely you know that you are God's temple" pleads St Paul, "that you are the place where the Spirit dwells."

Yes, Anglican Christianity is a sacramental religion. It certainly makes use of material things in its worship. It cannot get by without Eucharist or Baptism. But these activities can be celebrated in hospital wards and the homes of the elderly, in schools, indeed, anywhere. Pressured by lack of economic and human resources, Churches look radically at their needs and their histories. And a truly radical look, a grass roots vision of the essentials of Christian life, will, for many today, not only exclude the need for a dedicated church building, but see such a building as an obstacle to mission and an economic liability.

Encumbered (as they see it) by contemporary pressures to conserve and not to change, many congregations choose to vacate their expensive buildings and move instead into a hall. Some, in urban areas, disperse completely. Here in Manchester, a major church by the highly talented local 19th century architect JS Crowther is about to close. The stock of Crowther buildings has been severely reduced in recent years. There has also been the recent closure of a splendid 1920s church at Gorton by Walter Tapper, and the nearby highly majestic RC church by E W Pugin stands empty and crumbling. All of this is on top of the loss of Bernard Miller's spectacular 1930s Church at Withington, and follows a long history going back to the slum clearances of the 1960s where many fine 19th century churches were demolished wholesale.

Aside from this – but highly relevant – is the redevelopment of Owen Williams' 1930s Daily Express Building in central Manchester. Here, through glass, working printing presses were formerly seen by passers-by: work and life became one. But the presses are now gone. The building has been conserved and is used as office space, but robbed of its original function, has lost soul and character.

There is, however, a certain hope around. A growing trend in modern culture is the recognition that some are touched at a depth greater than words convey by atmosphere, by a sense of place, by a building and its aesthetic, its light and smell and history. Even those churches where the Word alone ruled supreme in the past are now using what they may call 'visuals'. To many of us these 'visuals' seem to be the natural consequence of recognising that Christian worship must appeal to all the senses if it is to be a valid experience of the incarnate God.

Such experience is not easily conjured in school halls and community centres. It requires the specific, the set apart. It is a very real part of the function of a building designed for use by the Christian assembly.

Yes, there are churches, many of them, where worship is still largely didactic and where the facilities of a seminar room take priority over the atmosphere of a shrine. But even within these churches, candles creep in on special occasions, flowers are used, and lights adjusted for special circumstances. Those minded towards conservation should encourage this trend. It points further than we think towards the discovery of common ground.

And perhaps, ultimately, that common ground – the Church as a place open to all, used by many, a place entirely without ownership – is the ground on which common understanding will be found.

Those responsible for our parish churches need to rediscover them as places for whole communities to use. They are indeed more than seminar rooms, even if a little less than shrines. Those who hear and spread the Word of God may need to search around that strange land where words fail and feelings take over. Equally, conservationists may need to move from an idealised view of history to a fresh understanding of growth and development. They may need to recognise that a church building, no matter how well conserved, if robbed of its use, loses soul and character in the same way as the Daily Express newspaper building I described earlier.

The plea then is for flexibility, wider understanding, and a conservation policy which allows church buildings to retain their biggest asset, which is to be the home of liturgy, not just for the regular worshipping community but for the entire parish. This will mean sacrifices by idealists on all sides. It certainly should mean better financial support – perhaps on the Continental model – by central government. And it might mean that local churches begin to look deeper at their worship and its conduct, as well as the wider and non-liturgical use their buildings can offer to their communities.

**NIGEL HAWLEY** is an Anglican priest and a former member of the Manchester DAC. His interest in Victorian architecture and conservation is finding expression in his present post as Rector of St Elisabeth's, Reddish, widely recognised as Alfred Waterhouse's finest church.

# THE CHURCHES OF ALEXANDER 'GREEK' THOMSON

St Vincent Street church ©Phil Sayer

## Jonathan Taylor

TASTE IS FICKLE. Each generation seems to react against the architecture of a previous generation, and what we consider to be our most important buildings today were often derided in the past, neglected and abandoned. The state of the architecture of Alexander 'Greek' Thomson and of his Caledonia Road church in particular typifies this reaction. Many of his most magnificent works, which were created in the mid 19th century, have been demolished or altered beyond recognition, and of his four churches, only one survives intact. Yet, in *The Pelican History of Architecture: 19th and 20th Centuries,* which was first published in 1958, the great American historian Henry-Russell Hitchcock described Thomson's three landmark churches in Glasgow as 'three of the finest Romantic Classical churches in the world', and today his work in that city is widely considered to be second only to that of the great Arts & Crafts architect, Charles Rennie Mackintosh.

Crown Copyright: Royal Commission on the Ancient and Historical Monuments of Scotland

**The interior of the Queen's Park church in the 1880s.** *The grille work in the centre at clerestory level hides organ pipes.*

©Glasgow Museums

**The Queen's Park church in 1869.** *This extraordinary creation of exotic details was bombed in the Second World War.*

## THE BIG THREE

The three landmark churches so admired by Hitchcock were Thomson's Caledonia Road church (completed 1857) which was gutted by arson in 1965, the St Vincent Street church (1859) which thankfully survives, and the Queen's Park church (1869) which was destroyed by an incendiary bomb in the Second World War. A fourth and much more modest church designed by Thomson at Ballater Street in Glasgow's Gorbals (1859) was substantially altered and then truncated by the construction of a railway arch in the late 19th century. This church was destroyed by arson in 1971.

The first two churches have strong similarities. The churches at Caledonia Road *(page 8)* and St Vincent Street *(page 5)* are both principally composed of a very tall, thin tower beside the much lower form of a Greek classical temple, raised up high above the street on a relatively plain podium. The juxtaposition of these two asymmetric elements on the podium is bizarre and highly dramatic, charged with tension, suggesting a place of ritual worship set apart from the world below – a theme echoed in the interior. The form was skilfully deceptive. In the case of the Caledonia Road church the main body of the church was set in the podium with the temple above in effect providing clerestory lighting. At St Vincent Street a steep slope enabled Thomson to take the effect one stage further; the church hall took up the whole of the podium with the body of the church above so that, from the south, the Greek temple form really was raised up by an entire storey.

Of the two, the Caledonia Road church is the simplest in detail externally. The church faces south with a fine Ionic portico on this elevation, fronting a much simpler structure of slightly different proportions. The tower, which is on the left when seen from the south, is monumentally simple, rising to three times the height of the body of the church, pierced by three slots to let out the sound of the bells and terminating in a cube of narrower section, capped with pantiles. Almost the only decoration is a cartouche on this top section and a simple incised Greek key pattern a little lower down.

The St Vincent Street church, which was completed two years after the Caledonia Road church, is much more elaborately ornamented externally. Where the tower of the Caledonia Road church is simply stylised, this tower terminates in a lavish floury of exotic detail which owes more to Asian and Indian architecture than to any Classical style, although much of the detail is Greek, such as the caryatid heads. The portico below it is, by comparison, a classic Greek temple, while the podium on which it sits is, in stark contrast to the tower, quite austere. Here, the plain stone walls are constructed with large ashlar blocks with narrow joints, and the only relief is the entrances which are dealt with in the heavily mannered style which typifies the architecture of the Greek Revival. The stone lintel over the Pitt Street entrance is, for example, heavily emphasised, turning it into a powerful, almost subterranean element by detailing it with a very heavy entablature supported on squat cube-like columns and by extending it on either side of the entrance over blind openings

These images of the Queen's Park church are taken from a digital model of the interior created by Sandy Kinghorn for an exhibition of Alexander Greek Thomson's architecture held at the Lighthouse in Glasgow in 1999. A full catalogue of the source material for the model, along with further images and animations can be seen at www.cadking.co.uk. This catalogue is a resource that anyone with material relating to Thomson can contribute to.

**Top left:** *A view of the interior from the gallery over the entrance, showing the reading desk raised high on a platform below the choir gallery and the organ, and framed by 'pylons' – Egyptianesque doorways with huge tapering surrounds*

**Top right:** *A detail of Thomson's elaborately decorated ceiling where it met the outer wall above the clerestory*

**Bottom left:** *The font in the centre of the church below the platform with the reading desk. The bowl of white Carrara marble was supplied by J & G Mossman*

**Bottom right:** *A view of the interior of the church at clerestory level looking east towards the gallery above the entrance*

**Centre:** *Three details which illustrate Thomson's startlingly original use of colour and pattern*

punched into the wall. Small, widely-spaced rosettes which decorate the lintel are a common motif of the Greek Revival, serving to establish a repetitive decorative element that emphasises the length of the lintel. Despite being flowers, the rosette motif is hard and crisp, and does not in any way soften the form, and the lack of a bottom moulding (the 'taenia' in the Corinthian order) ensures their dominance.

The most radical and the most exotic of the three landmark churches was the last, the Queen's Park church *(this page and opposite)*, which was commenced ten years later. Unlike its predecessors, the form was essentially symmetrical and the style primarily Egyptian in spirit. A strangely conical tower rose up above a central pediment supported on bulbous columns with papyrus heads, over a vast doorway framed by the tapering architrave that is known in Egyptian architecture as a 'pylon'. Photographs show that the main auditorium was even more lavishly decorated than that of the St Vincent Street church, with richly coloured bands of simple, stylised motifs, with the central pulpit flanked by huge pylons. According to the historian Gavin Stamp, "photographs suggest that the whole arrangement [of the pulpit] was much more like a sacrificial altar in a set by Cecil B DeMille for, say, *Cleopatra*. No one who saw inside this Presbyterian temple ever forgot it." He quotes Ford Madox Brown, the pre-Raphaelite painter, as saying shortly after its completion, "I want nothing better than the religion that produced this art like that."

## THOMSON'S STYLE

Thomson's clients for all three churches were the United Presbyterians. He was an elder of the Caledonia Road church congregation and his fervently religious approach to the design clearly found a receptive audience. The impact of the church architecture owes as much to his unique relationship with his client as to his own genius.

At a time when Gothic Revival dominated church architecture throughout Britain, Thomson derived his style of architecture from Greek and Egyptian sources. One justification given by the pre-eminent gothicist AWN Pugin and others for the Gothic Revival was that Classical architecture was the product of a pagan culture, and was therefore particularly unsuited to church architecture. However, Thomson and others took the line that, as a pre-Reformation style of architecture, Gothic may have been the natural choice for Roman Catholics, but not for Presbyterians. Furthermore, as the historian Sam McKinstry explains, Thomson believed that "God, in his Providence, revealed Himself [to Mankind] gradually through nature, then through the 'Elder Scriptures', and then through the Nations" – that is to say through the early civilisations. The architecture of Egypt and Greece thus reflected the glory of God as He had revealed Himself to them.

Although the choice of historical style was different, Thomson's approach to design was widely shared by the avant-garde movements of architecture and design at that time. For example, Owen Jones' highly influential *Grammar of Ornament*, published in 1856, is full of flat, stylised decorative patterns, richly coloured, and derived from many different periods and civilisations, including Greek and Egyptian in particular.

**The Caledonia Road church**, *the first of Thomson's three churches to be built (completed 1857), was gutted by arson in 1965. The tower has now been illuminated for the Crown Street Regeneration Project.*

## THE CALEDONIA ROAD CHURCH TODAY

When first built, the Caledonia Road church formed part of a cohesive townscape, contiguous with tenements by Thomson behind. This setting was damaged by the development of a railway viaduct which narrowly missed the western tenement in the late 1860s, and then destroyed by the demolition of the tenements in the early 1970s. As the economic prosperity of the area declined, particularly after the Second World War, the congregation could no longer afford to maintain their building properly. The congregation of the Caledonia Road church was finally dissolved in 1962 and the building passed first into the hands of the Church of Scotland and then to the Corporation of Glasgow. Now derelict, the condition of the fabric deteriorated rapidly. The interior was stripped of everything of any value and the rest was smashed or vandalised beyond repair. Finally, in October 1965, the vandals set fire to the building. The subsequent proposal to demolish the building was narrowly averted, partly due to the intervention of the architectural historian Henry-Russell Hitchcock himself. More recently, a lead roof has been added to the tower to stop the water pouring in, and floodlighting has transformed its appearance at night, but all the cement pointing and crumbling stonework has been left, and by day the building remains much as it did in 1965, a stark and highly dramatic, burnt out shell.

The future of the Caledonia Road church depends on finding a suitable new use for the structure, as there are still calls for its demolition – most recently from James Boyle, despite being the chairman of the Scottish Arts Council. With this in mind, the Crown Street Regeneration Trust organised a competition to find a new use for it, but its proposal has received a set back as it had hoped that the railway viaduct which isolates the building from others in the street might be demolished. This would have enabled the church to be reconnected with the rest of the street as it is, in effect, on a traffic island, making its reuse more difficult. But in August 2001 the transport authority determined to keep the viaduct in case it is needed in the future, despite the fact that it has not been used since 1966. Nevertheless, prospects for reusing the church look better than ever before as the regeneration of the surrounding area and the impending redevelopment of the streets around the building will bring new life back to the area, making a new use more viable.

## THE ST VINCENT STREET CHURCH

By comparison with the appalling treatment of the Caledonia Road church, the St Vincent Street church seems to have survived rather well. However, the setting of the church has been badly affected by the development of modern office blocks in the 1960s and '70s, and by the end of the 20th century its structure had deteriorated badly. The principal problems were; that it had been built of Giffnock sandstone which does not withstand weathering well, the use of cement pointing in repairs carried out in the 1960s, and its scale. Furthermore, like many inner city churches, demographic changes had also left the church with a tiny congregation which could not meet the cost of maintaining the building. The question is, who should? The principal authorities concerned had already shown a considerable degree of reticence to save one landmark Thomson building, despite intense lobbying from conservationists.

## ...THE WORLD MONUMENTS FUND INTERVENES

In 1998 the condition of the fabric and the lack of funding led to the St Vincent Street church being placed on the World Monuments Watch List. This list of the 100 most endangered sites world wide is maintained by the World Monuments Fund to help focus attention on buildings of international importance, usually where there is either a lack of will or a lack of resources within the country concerned to carry out the work required. The entry and the subsequent publicity brought international recognition for the importance of Thomson's work in Glasgow, and finally ensured the undivided attention of the national and local authorities. Work commenced on the most urgently needed work, the tower, two years later, with funding from the American Express Company, the Monument Trust, Historic Scotland, the Architectural Heritage Fund and the City of Glasgow, as well as a substantial contribution from an American donor, Mr Robert Wilson which matched private funding in the UK. The project was organised by the Glasgow Buildings Preservation Trust and directed by the Glasgow architects Page and Park.

## VALUE JUDGEMENTS

For the survival of Alexander 'Greek' Thomson's few remaining architectural gems we have to thank a small number of far-sighted and tenacious conservationists and architectural historians who recognised their value. Today it seems difficult to imagine how most people could have been so blind to the sheer quality of their architecture, their dramatic form and richness of detail as to allow their destruction. Yet even today, calls for the demolition of these most important Victorian buildings can come from the most unlikely sources and we have only to look at the plight of the Modern Movement architecture for a graphic illustration of the way conservation is still being led by a small number of avant-garde conservationists and historians, battling against ignorance and prejudice.

## RECOMMENDED READING


Stamp, Gavin and McKinstry, Sam (Eds); *'Greek' Thomson.* Edinburgh University Press, Edinburgh 1994

Stamp, Gavin; *Alexander Thomson – The Unknown Genius* Laurence & King, London 1999



This article has been prepared by **JONATHAN TAYLOR,** Editor, with the help of **Dr GAVIN STAMP**, Chairman of the Alexander Thomson Society, **IAN HAMILTON** of Page & Park Architects, and **TOM McCARTNEY** Director of the Crown Street Regeneration Project.


## THE WORLD MONUMENTS FUND

The WMF is the only private charity dedicated to the great monuments of the world. By raising awareness of the need to preserve the global architectural heritage, The World Monuments Fund acts as a catalyst for heritage conservation projects all over the world.

'Monuments' include palaces, fortifications, temples, churches, smaller buildings of architecturally significant types, and sometimes whole towns and settlements. Landscapes and gardens are also included. In Britain, the WMF prefers to take on sites in public ownership, and some public access is required in the restoration plans of all sites where WMF participates in the restoration.

The WMF is based in New York with offices worldwide. For further information contact:
The World Monuments Fund in Britain,
2 Grosvenor Gardens, London SW1W 0DH
Tel 020 7730 5344 Fax 020 7730 5355
E-mail wmf@wmf.org.uk
Website www.worldmonuments.org

©Phil Sayer

©Phil Sayer

©Page and Park Architects

©Page and Park Architects

©Page and Park Architects

**ST VINCENT STREET CHURCH** *(1859)*
*was placed on the list of 100 most endangered sites world wide maintained by the World Monuments Fund in 1998*

***Top:*** *The richly decorated interior of the church*
***Middle left:*** *The church from the east*

*Five photographs, taken by Page and Park architects, showing the work to the tower of the St Vincent Street church completed in 2001:*
***Centre:*** *Detail of one of the Caryatid heads close to the top of the tower. The lintel which almost rests on the head was skilfully removed and replaced, and the head was consolidated to slow down the rate of deterioration using an acetone/epoxy mix under the direction of Historic Scotland.*
***Middle right:*** *Detail of the pylon above the clock face showing replacement stone details. At the bottom of the picture is the elaborate sun burst motif which crowns the clock face, which is detailed inset.*
***Bottom left:*** *The top of the St Vincent Street tower, seen from the scaffolding during its recent repairs*
***Bottom right:*** *Lead flashings on one of the parapets, with acroteria*

©Page and Park Architects

# CAUTIOUS EXPRESSIONS OF FAITH

## Catholic Chapels in the Georgian Era

**Inside the 'Mausoleum' at Lulworth Castle, Dorset**

WHEN HENRY VIII cut England's ties to Rome during the Reformation the influence and integrity of the Catholic Church in England was very effectively broken. The Dissolution of the Monasteries and the pursuant destruction of the fabric of the Church was a savage and effective blow. The Catholic faith was left homeless and its congregation scattered and isolated as churches were appropriated for Protestantism. Fanaticism among the English Protestants was encouraged in order that England should be kept free from Catholic influence and subservience to papal authority.

During the 17th century a steady accumulation of laws eroded the freedom of worship and the human rights of all non-conformists in England. Those of Catholic faith were also targeted by specific laws. The *Acts of Uniformity* of 1545 onwards were passed in order to give English Protestantism a religious monopoly, binding state and church together. Religious diversity would make the monarchy vulnerable to faction and usurpation. These Acts essentially made failure to attend the established church illegal. Implicit in this crime was the assumption that the accused was not loyal to the monarch. As with Thomas More, the church and the state were inseparable; religion was not a private affair but was treated as a political statement. Being proved a Catholic was tantamount to treason. This was no political expedience: in the ensuing Elizabethan period and throughout the next century the throne was rarely free from attack. After the Restoration, the Catholics in England fared no better, as the balance of power had shifted to give Parliament more influence.

Political intrigue did not end until 1766 following the death of the final pretender to the throne, 'James III' to his Jacobite supporters. With his death also died the last threat of usurpation of the throne by the Stuart dynasty and Catholicism was freed from complicity with this political cause.

By the Georgian period harsh laws which ordered penalties for all aspects of Catholic life and worship were being enforced less and less. During the 17th century they had been used as a whip when necessary to quell any burgeoning confidence among the Catholic community. At the end of the century Catholics were still viewed with suspicion politically. Consequently they were unable to hold official posts and were excluded from William III's 1689 *Toleration Act* which gave non-conformist Protestants the legal right to build and attend places of worship.

Under this oppression the Roman Catholic Church was never able to assert itself and its illegal status gave its congregation and priests the mentality of outlaws. The practical elements of worship for small disparate

**The 'Mausoleum' at Lulworth Castle (John Tasker, 1776)** – *the first free-standing Roman Catholic place of worship to be built in England since the Reformation*

# THREE GEORGIAN CATHOLIC CHAPELS IN LANCASHIRE

**More Georgian Roman Catholic chapels are to be found here than in any other county**

**St Peter & St Paul, Stidd (near Ribchester), 1789:** *The wing on the right of this impressive farmhouse is a chapel. Only a bell distinguishes it from the other wing, a barn.*

**Hill Chapel, Gosenargh, 1802:** *Externally a simple chapel tacked on to the end of a row of cottages, like any other non-conformist chapel.*

**St William, Chipping, 1818:** *Before this chapel was constructed, services were held in an attic room of the farm house, and the signal to call the faithful to Mass was given by hanging out the washing on a line in the garden which could be seen for miles.*

*(All photographs by Caroline Hunter)*

congregations were dealt with in a discreet and even furtive manner, and a talent for disguise developed. Lest they drew attention to themselves, many congregations held low mass and even established permanent chapels in back rooms, outbuildings, stables and other unprepossessing buildings. By the 1770s however, Catholics had begun to feel more freedom to build, anticipating a near future when their faith would be emancipated. Their buildings over the next few decades though, were decidedly un-Catholic in style, vernacular and functional, or echoing the demure and simple neo-classical elements of their fellow non-conformists the Methodists and Baptists, reflecting the habit of mind that the beleaguered Catholic community had developed over the past 200 years; a disinclination to be noticed.

The establishment was taking a more tolerant view of the activities of Catholics, and George III himself is believed to have quietly sanctioned the building of an illegal Catholic chapel by the Weld family at Lulworth in Dorset, advising its real function be disguised as a mausoleum.

However, in 1780 those Catholics who had taken advantage of the relaxation and fulfilled the need of their congregations by building a public place of worship, were given a shock by the vehemence of anti-Catholic feelings that lingered among the public. The Gordon Riots in London were a violent expression of anti-Catholic feeling. A mob looted, burned and ruined not only Catholic chapels, but also the homes of Catholics and their sympathisers. It was a painful experience for English Catholics and as a result the style of their subsequent building was to be very restrained and unobtrusive. In 1791, the first *Catholic Relief Act* legalised the building of places of worship by Catholics. Despite this conciliation chapels that were built legally were still predominantly unassuming in style and built on inconspicuous sites. There were a few, though, who lamented the stylistic choices being made by those with a responsibility to build. The buildings of Joseph Ireland, for example, anticipated the glory and splendour of the Gothic Revival and the work of Pugin. By the Regency period and the time of the final *Catholic Emancipation Act* of 1829 the Catholics of England had begun to recover and reinterpret their built heritage.

## ST BENET'S CHAPEL AND PRESBYTERY, NETHERTON

St Benet's at Netherton in Sefton, on the north side of Liverpool, is a classic example of the low key vernacular chapels built by Catholics in the late 18th century. (Curiously, there are more early Post Reformation Catholic chapels in Lancashire than any other county.) Built in 1793, the chapel is tucked lengthways behind its presbytery, the two being integrated internally, so that it looks like an ordinary house from the street, reflecting the anxieties and restrictions imposed on Catholics following the Catholic Relief Act of 1791. It is now in the ownership of The Historic Chapels Trust (HCT)[1] and is listed Grade II*.

The building exudes a simple period charm, in brick with stone dressings. The chapel roof is covered with stone slates, that of the presbytery is slated. Its frontage has a well-crafted, pedimented door-case and fanlight. Until the 1970s the chapel interior remained largely intact but later the seating and various furnishings were removed when the building was no longer used for worship. HCT is now carrying out a careful repair of the chapel, inside and out. Fortunately the gallery with its original joinery, gallery railings and cast iron supporting columns still survives. Most important of all is the delicate plaster altarpiece set in a pilastered pedimented frame. The central panel is painted in colours reminiscent of the dawn sky, with a relief of swagged curtains, at the top is a descending dove and a rising sun. According to Bryan Little, "With its winged cherub heads and gloria of rays and Adamesque urns and garlands, [this composition] is of the type that many churches of the Establishment could boast before the zealous efforts of 'ecclesiological' restorers" (Little, 1996, see *Recommended Reading*).

The sarcophagus altar probably dates from the early 19th century and is placed behind altar rails hard against the east wall, thus it predates the general layout of Catholic churches prescribed by the Second Vatican Council. In the lower part of the Georgian sash windows are examples of rare 1920s glass 'transfers' whose conservation is now being undertaken.

The parish of St Benet is one of the oldest in Lancashire. Before the Reformation, worship centred on St Helen's church in Sefton (Grade I) where monuments to the Molyneux family can be seen. They remained loyal to the Catholic faith through the Reformation. Catholics worshipped in their house at Sefton Hall, where in 1603 a Benedictine priest is known to have been officiating. Eventually a separate place of worship was established in a barn at Netherton that served as a church from 1769-93. The building of St Benet's Chapel was commenced in June 1792 and it opened in 1793. It was made redundant in the 1970s when a larger church was constructed nearby, and the Historic Chapels Trust acquired both chapel and presbytery in 1995. HCT is repairing the chapel and will eventually be reinstating some of its furnishings. The trustees hope that St Benet's will become popular for community activities and occasional services of worship, especially weddings.

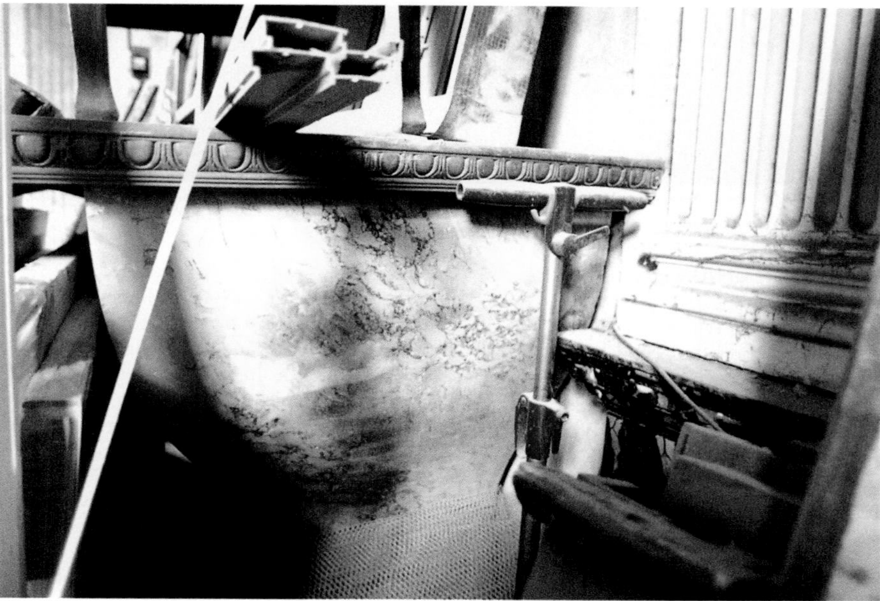

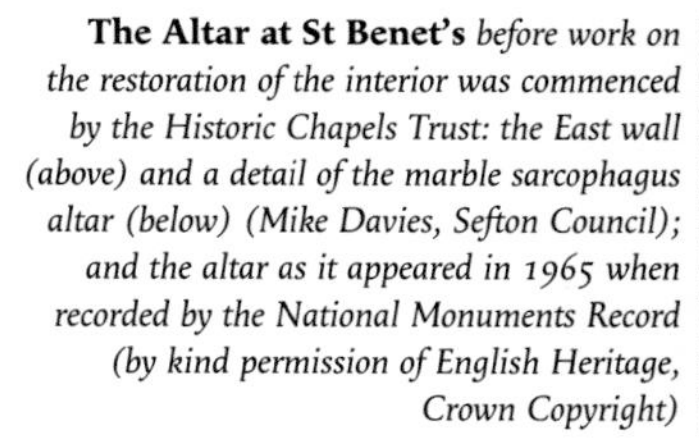

**The Altar at St Benet's** *before work on the restoration of the interior was commenced by the Historic Chapels Trust: the East wall (above) and a detail of the marble sarcophagus altar (below) (Mike Davies, Sefton Council); and the altar as it appeared in 1965 when recorded by the National Monuments Record (by kind permission of English Heritage, Crown Copyright)*

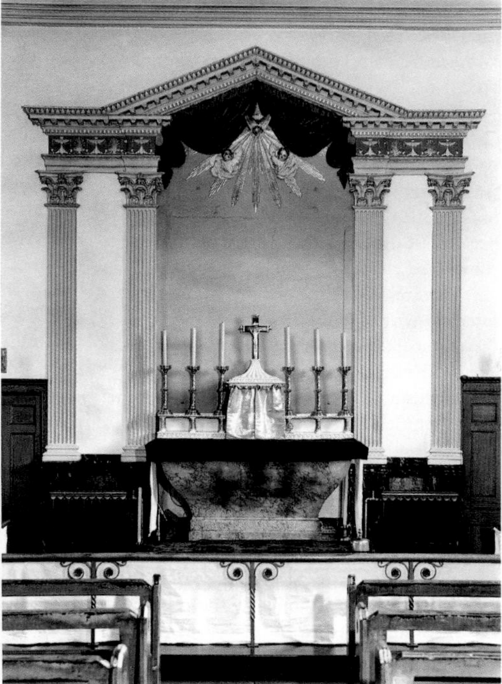

***St Benet's, Netherton, Merseyside, 1789;*** *the chapel is discreetly tucked behind its presbytery and is barely visible from the road (Historic Chapels Trust)*

**New Wardour, Wiltshire (James Paine, 1776)** – *the chapel is in the wing on the left, and is not in any way distinguishable externally*

## WARDOUR CHAPEL, WILTSHIRE

Built between 1770 and 1776 by James Paine, New Wardour in Wiltshire is a splendid Palladian mansion. It was built by the 8th Lord Arundell to replace Old Wardour Castle which was destroyed during the Civil War by its owner in order to gain the surrender of the Parliamentarian troops who had captured it.

By the late 18th century the Arundells were the leading lights of the Catholic congregation in the Salisbury area. In 1780 when Fanny Burney visited Salisbury, she noted that "There is no Romish chapel in the town; mass has always been performed for the Catholics of the place at a Mrs Arundell's in the Close – a relation of his Lordship." (From *Wardour, A Short History* by Philip Caraman, 1984.)

By this period there were 40 to 50 Catholics in Salisbury and 540 in the combined parishes of Tisbury, Semley and Ansty. It was thought to be the largest congregation of Catholic recusants outside London, partly because of the attraction of the recently completed chapel.

The architect James Paine was given a free rein with the building of the house at New Wardour. However, Lord Arundell was keen to be very involved in the design and construction of his, ostensibly, private chapel. Lord Arundell used Fr John Thorpe, an English Jesuit based in Rome to liase with Italian workmen and artists including another Palladian architect called Quarenghi. What resulted was a glorious and ostentatious Catholic chapel, unusual for its time but nevertheless an expression of the growing freedom given to Catholics by the establishment. Lord Arundell was in very regular contact with Fr Thorpe, discussing the designs and asking for advice about the furnishings. Quarenghi was given responsibility for the chapel's interior which was furnished magnificently, with large wall paintings; the altar, unusually placed in the West end, was made from a variety of expensive marbles by Quirenza to a design by Quarenghi; the gilded sanctuary lamps were made by Luis Valadier and, thanks to a nifty piece of architectural salvage on Fr Thorpe's part, the altarpiece came from the private chapel of the Jesuit Superior General after the suppression of the Jesuits.

After completion in 1776, (15 years before it became legal for Catholics to build chapels) the chapel which was, as Nikolaus Pevsner notes in *The Buildings of England*, actually the size of a small parish church, was opened with great ceremony by Bishop Walmesley and was used by tenants, retainers and locals. The size of the congregation was noted, but tolerated by the local Anglican administration. The vicar of Semley in 1780 sent his bishop a return for his parish with this apology; "Should the number of Papists seem large for this parish, which is not a populous one, your Lordship will easily account for it from its vicinity to Wardour Castle." (From *Wardour, A Short History* by Philip Caraman, 1984.)

By this time the chapel had been serving the local community for four years and, after the first *Catholic Relief Act* in 1778, was a famous centre for Catholic worship in the West. Because of its high profile it became a target in 1780 during the Gordon Riots and was nearly burnt down. However, it survived this threat, but not without changes; Lord Arundell for years after the Riots, employed a keeper of the peace to attend the masses at the chapel. The crude pew that was installed for this man is still in place today.

In 1789 John Soane visited Wardour, and soon after was commissioned to enlarge the chapel. He is responsible for the chapel's dome, oval vault and a shallow apse. Despite the threat of anti-Catholic feeling the chapel, and by implication, the congregation grew and survived. In fact by the time of the *Catholic Emancipation Act* of 1829, there were two catholic schools in the area and Lord Arundell had given refuge to a French bishop escaping the Revolution and settled a small community of monks driven from their home in Bourbon.

## RECOMMENDED READING


Little, Bryan, *Catholic Churches since 1623*, Robert Hale, 1996, p44

Cross, F L (ed), *The Oxford Dictionary of the Christian Church*, OUP, 1974

Parker, Derek and Chandler, John, *Wiltshire Churches, An Illustrated Guide*, Alan Sutton, 1993

Roberts, Jane, Tranmar, John; Cussans, Anthony, *Wardour Castle: A Guide to the House, Chapel and Grounds*, for Cranbourne Chase School, 1976

Anthony Williams, J, *Catholic Recusancy in Wiltshire*, Catholic Record Society, 1968

Oliver, George, *A History of the Catholic Religion*, Charles Dolman, 1857

Davies, Norman, *The Isles*, Macmillan, 1999



This article was prepared by **HANNAH MOFFAT** of Cathedral Communications and **Dr JENNIFER M FREEMAN** of the Historic Chapels Trust contributed the section on St Benet's.


[1] The Historic Chapels Trust was established in 1993 to take disused chapels and other places of worship in England into ownership. Its buildings are all of outstanding architectural and historic interest. The Trust undertakes their preservation and repair for public benefit and opens its chapels to visitors and for a variety of suitable activities.

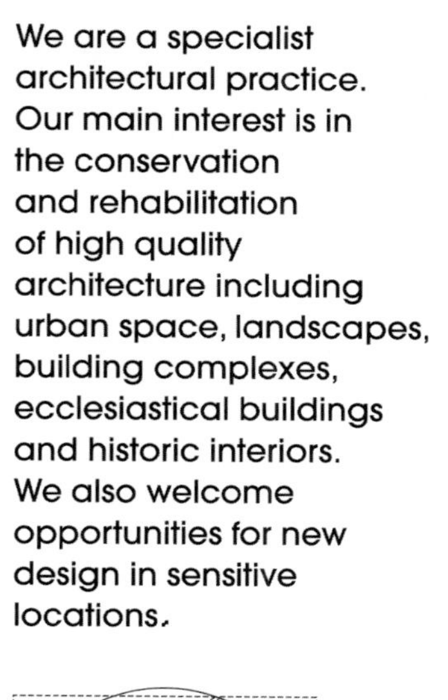



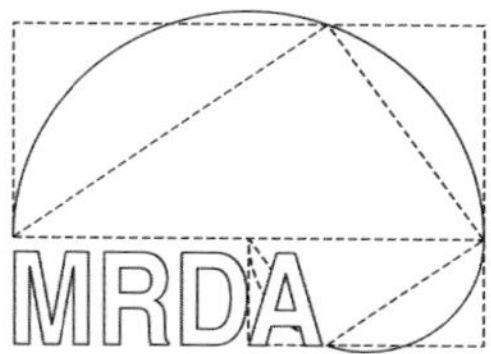
MRDA

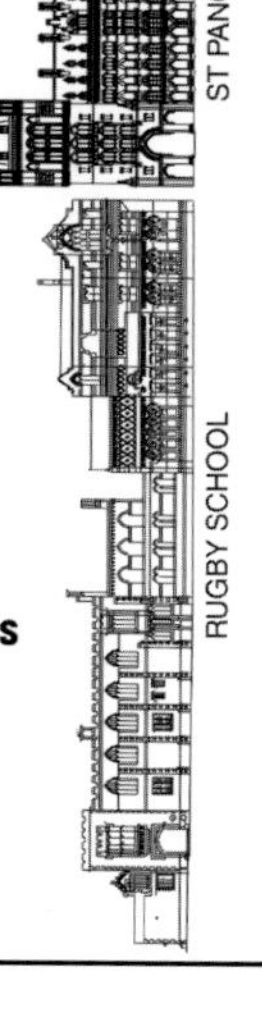
TRURO CATHEDRAL
ST PANCRAS CHAMBER
RUGBY SCHOOL



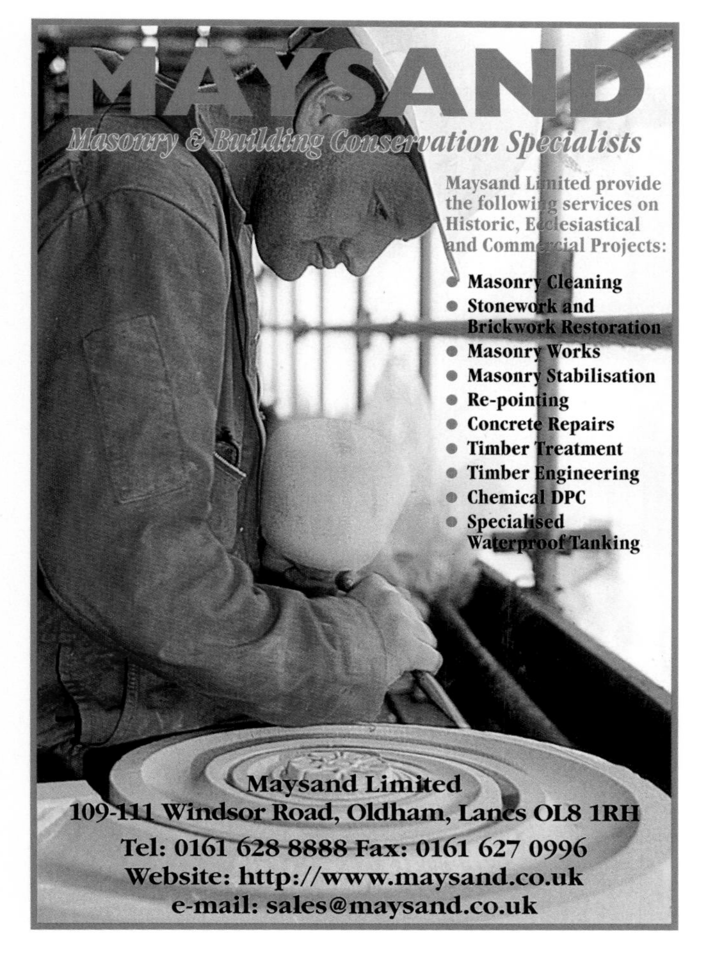
MAYSAND
Masonry & Building Conservation Specialists
Maysand Limited provide the following services on Historic, Ecclesiastical and Commercial Projects:
Masonry Cleaning
Stonework and Brickwork Restoration
Masonry Works
Masonry Stabilisation
Re-pointing
Concrete Repairs
Timber Treatment
Timber Engineering
Chemical DPC
Specialised Waterproof Tanking
Maysand Limited
109-111 Windsor Road, Oldham, Lancs OL8 1RH
Tel: 0161 628 8888 Fax: 0161 627 0996
Website: http://www.maysand.co.uk
e-mail: sales@maysand.co.uk

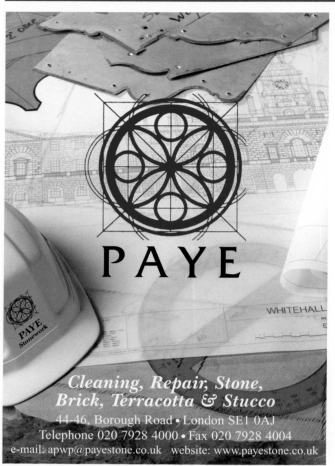

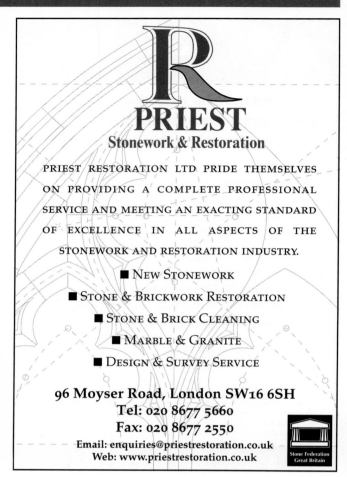

# THE CONSERVATION OF ALABASTER

*A detail of a 17th Century alabaster effigy in Bath Abbey*

## Dennis Cox

ALABASTER IS A FORM OF GYPSUM, hydrated calcium sulphate. Gypsum is used extensively in plaster and mortar and is added to Portland cement to inhibit the setting period. During quarrying or mining for this mineral, large compounded blocks would be encountered and these were alabaster. Unfortunately, according to the technical advisor to the Stone Federation Great Britain, John Bysouth, alabaster is no longer extracted in the UK, or if it is it isn't appearing on any major radar screen. A British Gypsum spokesperson said that as far as their estates are concerned it simply doesn't exist in large enough quantities to make its continued extraction economically viable. However, there is still a significant mined reserve of British alabaster left in the UK. It is owned by the Nigel Owen Organisation of Northampton. This company, which now quarries soap stone, finds itself the owner of a spectacular 'bin-end' of what is probably the last of the UK's stock above ground. Nigel Owen himself must be the UK's leading expert on this stone having been involved with alabaster since 1945. He reckons that at the present rate of take up (mostly to sculptors, who won't thank me for divulging this information) there is a two year supply. It is available in workman-like blocks of around a foot square.

These sculptors who are buying from the Owen bin-end represent the last knocking, or more properly, tappings of a great British tradition that, according to most authorities, really took off in the late Middle Ages, just before the Black Death.

Far too soft for external use, Alabaster is hopelessly vulnerable to the elements, and it is too weak for building. It is so soft that it can be cut with a penknife, making it a ready target for vandals over the ages. The very ease with which it could be carved with finally wrought detail was one of the reasons for its popularity for tomb sculptures and other internal devotional features, like reredoses, triptychs and panels. The other reason, of course, was its spectacular appearance. All alabaster has a unique translucent quality that even the most desensitised individual should be able to distinguish from marble. The colour spectrum ranges from a creamy white (very rare) to a dark honey colour. Whatever the colour, it is made up of dozens of veins. In colour these may be anything from dozens of shades of white to veins of pink and reddish brown.

The discovery of large accessible and workable deposits of alabaster around Nottingham at the end of the 14th century arrived just in time to satisfy the aspirations of all the *arrivistes* who had prospered in the period that followed the Black Death. They wanted to both demonstrate their gentility in the here-and-now and guarantee their place in the hereafter via an ostentatious contribution to their local church. Easy to carve, easy to work, British alabaster was the ideal stone to cope with needs of *the mass affluent*. Production from Nottingham reached the sort of proportions that were not to be experienced again until the 18th century in the potteries. It didn't just stay in England. English alabaster, from small votive statues to the Virgin to large funerary monuments and tombs, was exported from Nottingham all over European Christendom, from Iceland to Spain. It has to be said that the quality of much of this production was very poor, often knocked out in pattern book style, seemingly by the yard. Late Medieval and early modern alabaster does have its great moments, however, and most of it is still highly acceptable when compared to the great Victorian alabaster carving revival when, what Alec Clifton-Taylor memorably referred to as a 'streaky bacon' alabaster, stalked the land along with other strange Victorian ecclesiastical enthusiasms. Alabaster is not just confined to churches however, and its use in the so-called 'Marble Hall' of Holkham Hall, Norfolk is regarded by some authorities as the way that it should be used to create a spectacular, almost theatrical effect. However, most alabaster encountered by the restorer and conservator is located in churches, more often than not as part of a funerary monument.

### CONSERVATION AND REPAIR

There are certain rules of engagement that apply to alabaster, as they do to any stone that is undergoing conservation. Firstly, whatever procedure is employed must, wherever possible, be reversible and secondly, the aim of the exercise is not costume drama, attempting to return the subject to its original form, but rather to stabilise the subject in the present. Fortunately that means the virtual

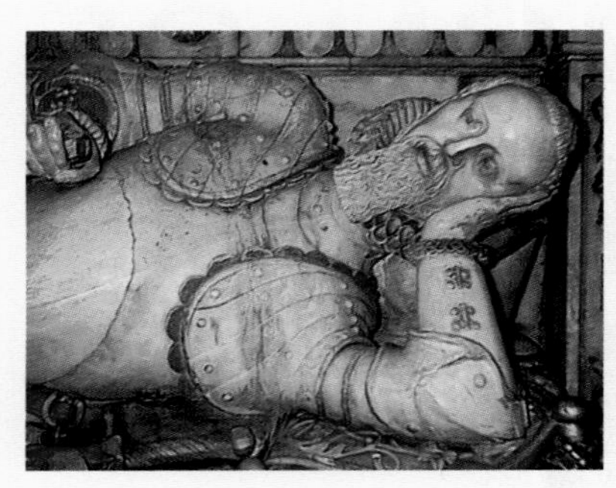

***The effigy of Sir Raphe Weldon*** *(1609) at St Peter & St Paul's, Swanscombe, Kent during conservation*

***The effigy of Dame Ann Carew*** *(1605) from St Edward the Confessor's, Romford, in the studio of Taylor Pearce after cleaning and repair. This effigy is now re-fixed on the restored monument in the church.*

***Sir Augustine Nicoll's effigy*** *being examined in the studio of Taylor Pearce for the Victoria & Albert Museum. The effigy was subsequently installed by Taylor Pearce in the new British Galleries.*

***St Margaret, Barking****: detail of relief part of the monument to Sir Charles Montagu (1625)*

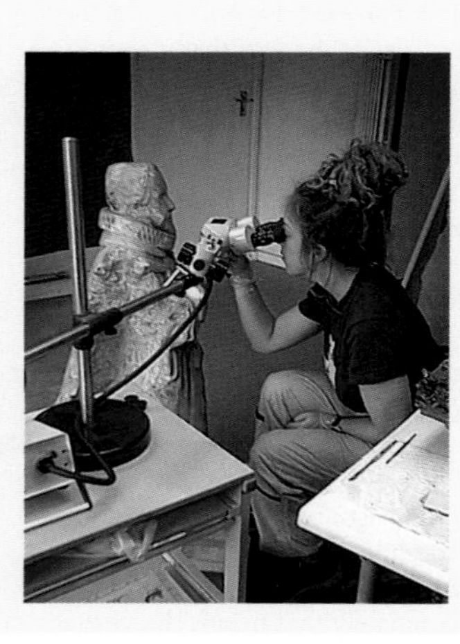

***Detail of monument to Thomas Withering*** *(1625) being cleaned in the studio of Taylor Pearce. Thomas Withering was Postmaster General to Charles I.*

disappearance of alabaster from the scene is not a problem from the conservator's point of view. Missing heads and extremities do not need to be replaced – that is provided they were lost as the result of several hundred years of wear and tear or even removed by a vandal, if it was a good class of vandal and a long time ago, like one of Edward VI's or Cromwell's iconoclasts. However, the activities of modern vandals have to be expunged from the historic record.

All vandalism must be left to specialists to treat. Spray paint for example may be readily removed using acetone or dichloromethane, but tests may be necessary to establish which solvent is required, and skill is required to remove the pigment thoroughly.

Before any work can be undertaken it is absolutely necessary that a condition report and a specification of work be prepared. As often as not all the work required cannot be undertaken on site, and the subject, or portions of it, have to be dismantled and removed to the conservation studio or workshop. Anybody budgeting for an exercise of this kind should bear in mind that a dismantling process may also reveal other problems hitherto unsuspected, that may have to be tackled. Recently conservators from Taylor Pearce encountered a nest of wild bees in a tomb. Clearly aware of their protected status the bees refused all the blandishments of an apiarist to resettle them, and in the end the conservators managed, with the agreement of the client, to strike up a *modus vivendi* with them, managing to work around them.

Funerary monuments account for the bulk of alabaster encountered by the conservator. However, few of these consist entirely of this material. It is often used in combination with marbles and sandstones, for instance, which require similar but different conservation procedures.

The main problems that alabaster presents to the conservator spring from its softness and susceptibility to the elements, and one of the most common, caused by water penetration, is rusting ferrous fixings and cramps. Besides staining, corroding iron can cause the stone to fracture and ultimately render a complete monument unstable and dangerous. The treatment is to remove all ferrous fixings and replace them with stainless steel fixings and cramps set in resin adhesives. Sometimes the troublesome ferrous material was introduced in 19th century restorations, as the fixings of many monuments down to the 17th century were more usually non-ferrous. Sheep bones were a particularly popular alternative. Indeed probably one of the biggest problems facing the conservator of alabaster, or any other stone, is botched or failed restoration procedures carried out in the past. Even the first epoxy resin repairs are now starting to emerge as subjects for treatment.

Much can be done to correct poor restoration or repair. For instance old repair plaster can be removed from a feature and break joints filled with a matching aggregate combined with a resin fill such as Paraloid, touched in to match the adjacent surface.

The area where alabaster needs the greatest attention, and where the results of a job well done are particularly gratifying, is in the cleaning and finishing processes. It may seem unecessary to observe that this is not a job for amateurs, but there have been a number of well recorded atrocities carried out, particularly against alabaster funerary monuments, by well meaning incumbents and their flocks over the years.

Usually alabaster, however begrimed by the centuries, will clean up beautifully. One great virtue of this stone is that, unlike marble, it rarely stains and if it does, as often as not the stain can pass off as one of its veins. In fact the veins are a kind of staining.

It should be remembered that up until the beginning of the 17th century, alabaster monuments were usually painted in places. Therefore it is important to carry out a careful examination for traces of polychromy so that paint layers can be tested for stability and resistance to solvents prior to cleaning. Subsequently any flaking areas discovered during cleaning need to be consolidated with Paraloid resin.

Alabaster should be cleaned using mild solvents. The type and strength of solvents has to be determined by on-site tests but the most usual is a mix containing white spirit, de-ionised water and a small quantity of non-ionic detergent. The actual cleaning is carried out using cotton wool swabs dampened by the solvent. After cleaning the subject should be protected by the application of a cosmolloid wax. This will restore the magic quality to this most agreeable of stones.

**DENNIS COX** is the Director of Taylor Pearce Restoration Services Limited, a firm of conservators specialising in the conservation and restoration of stone statuary and ornaments, architectural ceramics, mosaic work and church monuments. Tel 020 8297 1599.

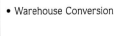

# 'BUT THEN WE SHALL SEE FACE TO FACE'

*(1 Corinthians 13 v12)*


## Susan Mathews

***Nicholas Ridley**, second half of the 16th century.*
*Believed to be from Leez Priory, Essex*

FOUR PORTRAITS in stained glass in the collection of the Stained Glass Museum at Ely Cathedral provide a fascinating insight into the importance of stained glass windows as works of art and the development of the craft.

The Stained Glass Museum, located in the magnificent setting of Ely Cathedral, is dedicated to British stained glass of all periods. Amongst the many stained glass panels which make up its permanent exhibition are a number of portraits with a higher than average degree of realism. Two of these, of Nicholas Ridley and George III, are depicted in costumes and settings that immediately convey their milieu. That of Nicholas Ridley dates from the second half of the 16th century. It is a small panel, painted on a single piece of white glass using a variety of enamel pigments in the painstaking manner of a miniature. Although by the time this was painted, the art of portraiture, as paintings, had begun to spread in England, the art was not yet as sophisticated as it was in, for example, Italy, and was confined to the higher ranks of society. As David Piper put it[1], "Bishops showed a worldly interest in perpetuating their mortal images leaving behind them a series of gaunt, wooden and unfriendly faces". Indeed the panel depicting Nicholas Ridley is a rather severe little piece, devoid of the brilliance and rich colour of earlier medieval glass. Bishop Ridley faces three-quarters right and his gaze is directed straight ahead as if he is actually sitting for the portrait. He is bare-headed and dressed in simple, unadorned black and white, reflecting the sobriety of a protestant reformer. A plain amber band encircles the portrait, beyond which, in the four corners, stickwork – achieved by scratching with a stick or needle into unfired paint – produces a decorative effect. The glass is painted on both surfaces. This was quite a common practice from early medieval times and was employed to create a special effect.

At Tattershall in Lincolnshire, a 15th century Head of a Woman, has the details of her face painted on the inside of the glass and her veil on the outside which emphasizes the floating nature of the fabric. In the east window of Holy Trinity, Goodramgate, York, is a 15th century figure of St Peter standing in water. The feet are painted on the inside, and fish and waves on the outside emphasizing the depth of the water.

In this portrait of Nicholas Ridley, the details of the face are painted on the front and the shading of the beard and hair is on the back. The pallor of the face is relieved by a tint known as sanguine, also applied to the reverse side of the glass. Sanguine and carnation were coloured pigments used for flesh tones and lip colours. The panel is set within later coloured glass of the 18th century.

[1] *David, Piper 'The Painted Face' The Elizabethans – A Rock Against Time, p 58 London, 1978*

Enamel pigments began to be used in the late 15th century. During the 16th century a wider range was developed and included blue, lilac and pink. These coloured enamel paints enabled stained glass artists to achieve effects hitherto impossible. They could be applied to white (or other coloured glass) in the manner of painting on china, in that a number of different colours could be applied together on the same piece of glass. They were used in conjunction with 'silver stain', available to glass-painters since the early 14th century and used to stain white glass yellow. Enamels increased the realism and naturalistic effects that could be achieved in glass painting although they lacked the lusciousness of a window made of coloured glass. Enamels were used ever more extensively during the 17th and 18th centuries as coloured glass was in increasingly short supply and because architectural style and artistic tastes were changing.

In the 18th century a number of famous painters were attracted to the medium of stained glass. Sir Joshua Reynolds for example, was invited to submit a design for a new window in New College, Oxford. This was then executed as a stained glass window between 1779-83 by the Dublin-born glass-painter Thomas Jervais. It depicts the Nativity, with female figures of the Christian virtues above. Reynolds, first president of the Royal Academy, was renowned as a portraitist. Famous society beauties posed for the virtues, earning the scorn of those who christened the New College figures 'Sir Joshua's harlots'!

This second portrait is of George III. The king dressed in his coronation regalia and seated in the medieval coronation chair is on loan to the Stained Glass Museum from the Royal Collection. It is a faithful copy of one of a pair of royal portraits by Sir Joshua Reynolds, painted in 1780. Many contemporary copies were made and here it is interpreted in glass by the stained glass artist James Pearson in 1793. It is extremely large (2.4 x 1.5m), dwarfing the Ridley panel not only in size but in execution also. Pearson has achieved astonishing effects – the sheen on the stockings, the glossiness of the jewels in the garter collar and the individual hairs of the ermine are all meticulously painted and extremely realistic. Visitors to the Museum find it an arresting exhibit, quite outside their expectations of a stained glass window. It would have been quite a show stopper in its day and was 'made to impress', rather than with a specific location in mind.

Portraits such as these were put on exhibition in London and were displayed in a darkened room with artificial lighting. The public were invited to admire at close quarters the virtuosity of the artist's ability in committing an original work of art to glass. Sadly for Pearson, the panel did not find a buyer. Although George III was a great patron of the arts and of large-scale contemporary glass-painting, public knowledge of his mental instability and a degree of royal unpopularity, plus the high cost of Pearson's windows left the panel unsold. It eventually found its way to the State Apartments at Windsor where it was displayed in the King's Dining Room above a chimneypiece and illuminated by candles. By 1902 it had been removed as it was no longer

*George III, by Joshua Reynolds and James Pearson, 1793. On loan from Her Majesty the Queen.*

***The Duke of Clarence as St George,*** *by CE Kempe & Co, 1905, from the Ambassador's Staircase, Buckingham Palace. On loan from Her Majesty the Queen.*

***'Self-Portrait IV'*** *by Debora Coombs, 1994*

felt to be in keeping with the décor. The panel did not see the light of day again until 1990, when it was rediscovered languishing in a store, wrapped in newspaper dated 1912. It was lent to the Stained Glass Museum and restored in 1991 by Chapel Studio of Hertfordshire.

The third stained glass portrait is of the Duke of Clarence in the guise of St George, commissioned by Queen Alexandra after the death of Albert Victor, Duke of Clarence, heir apparent, in 1892. It is also on loan to the Stained Glass Museum from the Royal Collection. Made in 1905 by CE Kempe & Co, it originally formed part of a large set of windows on the Ambassador's Staircase in Buckingham Palace and was Kempe's most prestigious domestic commission. The Palace was bombed during the war and much of the glass destroyed.

The figure is magnificently attired in golden armour and crowned with both a halo and a laurel wreath. The face however, in photographic sepia, is startlingly realistic and somewhat at odds with this flamboyant figure with legs akimbo astride a shrinking dragon.

As with all the work of the Kempe studio, it is extremely accomplished in execution. A restricted palette of white, blue, green and red, greatly enriched with silver stain is typical of work drawing on the style of late 15th and early 16th century English stained glass.

The output of the Kempe studio was prodigious, with examples in most cathedrals and parish churches in Britain and many in the United States. Kempe's work is very distinctive in style and can often be identified by a maker's mark based on a wheatsheaf, usually painted in a lower corner. A loyal and devoted body of enthusiasts formed The Kempe Society in 1987, the 150th anniversary of his birth. The firm closed in 1934.

The fourth portrait in glass entitled 'Self-Portrait IV', is by the artist Debora Coombs. The paint has been applied quickly using fingers, sticks and brushes. The result is a refreshing piece of work that has an immediacy not often achieved in stained glass. Stained glass artists use hazardous materials and the craft is very labour intensive which can sometimes result in pieces of work that are accomplished but lacking in spontaneity. This panel defies these restraints. The glass used here is not precious, mouth-blown glass, but has been selected from pieces that were to hand and perhaps from the scrap box: one is patterned and acid-etched – the sort used in bathroom doors to afford a degree of privacy. The panel was one of nine, made for the exhibition One Woman's Narrative, for the Cochrane Theatre, London in 1994 and is intended to convey the pain and conflict of modern womanhood.

The artist now lives and works in the United States and has recently completed a commission in St Mary's Cathedral, Portland, Oregon. The commission was for 20 windows, nine of which were stained glass portraits of American saints.

The portraits in the Stained Glass Museum have proved endlessly fascinating for visitors, especially children. They have a feeling of immediacy lacking in the many figures of saints and angels. There are many other intriguing images which might be portraits, but which are not identified. They are most certainly taken from life using studio models who will now forever remain anonymous.

## RECOMMENDED READING


Brown, Sarah and O'Connor, David *Medieval Craftsmen – Glass Painters,* London 1991

Stavridi, Margaret, *Master of Glass: Charles Eamer Kempe 1837–1907,* Hatfield 1988

Baylis, Sarah, *'Absolute Magic' A Portrait of George III on Glass by James Pearson, The Journal of Stained Glass* Volume XX II 1998 pps 16–30

Clarke, John, Ed Antonia Fraser *The Life and Times of George III,* Weidenfeld & Nicholson 1972



**SUSAN MATHEWS** has been the curator of the Stained Glass Museum at Ely since 1990. Previous to that she had her own stained glass workshop in Leyland and Cambridge. From 1969–1983 she taught in high schools in Buxton, Haywards Heath and Lowestoft.

# DOING THE CHORES

## Interior maintenance for historic churches

Jonathan Taylor

*Photograph by kind permission of Bath Abbey*

*Left: Vacuum cleaners lift dirt from the cracks far more effectively than a broom Right: Dusting with a soft brush is safer and more effective than using a duster, reaching in between mouldings and into the crevices of the carving without snagging*

Owners of historic buildings may find interest, even excitement, in the prospects of a conservation scheme that will enhance their building and realise its character, beauty and historic interest. But few people are going to get excited about cleaning out the gutters and sweeping the floor. You may revel in the skill of a fine craftsmen, but who appreciates the skill of the cleaner? So, it should not be surprising if the cleaner takes little pride in his or her work, and that it is the maintenance budget which gets cut first. But it is this attitude which leads to disaster. According to Timothy Cantell, of Maintain our Heritage, more historic building problems are caused by neglecting maintenance work than any other cause.

No matter how mundane these tasks may seem at face value, each requires careful thought and consideration where the fabric of an historic building is concerned. Each needs to be considered holistically, in the context of how the building ages and decays, and how every action may affect the deterioration process. For example, the simple, ordinary task of dusting, which everyone takes for granted, often simply moves the dirt around on the surface like sandpaper, and over the years gentle dusting with a dirty duster gradually abrades the surface beneath it.

### KEEPING OUT THE DAMP

*"Regular maintenance and repair are the key to the preservation of historic buildings. Modest expenditure on repairs keeps a building weathertight, and routine maintenance (especially roof repairs and regular clearance of gutters and downpipes) can prevent much more expensive work becoming necessary at a later date."*

*PPG 15 'Planning and the Historical Environment'*

Water can enter a building from rain through leaking roofs, walls and windows, from ground water, and from leaking taps and drains. Valley roofs, gutters, hoppers and downpipes need to be kept clear of leaves and dead pigeons and must be cleared in the autumn after the leaves have fallen. Leaks should be noted and reported immediately, so that repairs can be made before they cause extensive damage.

Materials may also be affected by condensation and by the water introduced when cleaning – bear in mind that a bucket and mop will saturate porous materials and even the water introduced when cleaning a surface may cause considerable damage to a vulnerable finish such as gilding. All these sources of damp need to be considered by those responsible for maintenance.

Organic materials such as wood and textiles decay in a damp environment due to organisms like dry rot (a fungus) and the larvae of certain beetles which feed off the organic material and flourish when the conditions are right for them. Inorganic materials may also decay in damp conditions. Stone may be harmed by the movement of salts which crystallise within the pores of the masonry. Interior gilding is particularly vulnerable, and even glass is damaged by moisture.

Changes in relative humidity (RH) affect organic materials as these retain moisture in variable amounts, expanding and contracting with variations in relative humidity.

Composite materials such as paint on wood may absorb moisture to varying degrees, causing the two materials to separate, or a paint layer to flake. At moisture levels above 65 per cent RH, mould may start to grow; below 50 per cent cracking of wood may occur.

## HEATING

It would be nice if a church could be heated instantly, just before the congregation arrives, without harming the fabric of the building. Unfortunately, sudden increase in temperature cause materials to expand rapidly. Humidity levels also change rapidly, causing condensation on cold surfaces, and studies have shown that intermittently heated churches are at greater risk from timber decay than those which are not heated.

To minimise the risk of condensation, the National Trust aims to keep its buildings at a temperature of 5°C above the outside temperature, and heats them throughout the winter even though they are shut. Ideally, a church in winter should be maintained at about 10°C, assuming an average exterior temperature of 5°C, but few congregations are going to find this acceptable in winter. Robin Wright of Lightwright Associates, a specialist services engineer, recommends increasing this temperature to 18°C for the period of the service – a level which a warmly dressed congregation will find tolerable. Keeping the church gently heated (at about 10°C) the rest of the time is likely to increase heating bills

*Weeds growing out of a hopper indicates a blockage which will inevitably lead to far more serious problems inside the building*

by just ten per cent, and involves much less strain on the fabric of the building, reducing long term maintenance costs.

Ventilation is also important, particularly in closed roof spaces, to maintain a low risk of condensation.

## SUNLIGHT

Most historic churches are relatively dark inside, so sunlight is rarely a problem for most church fittings and furnishings. However, hanging banners which are close to a window are often vulnerable. Ultra-violet light in sunlight causes textiles to oxidise and the fibres become brittle. Timber may also be bleached by the sunlight and dry out, causing the surface to crack.

## CLEANING

The day-to-day maintenance regimes include the removal of dirt not only to keep the building clean, but also to prevent wear and tear. Floors are particularly vulnerable to the grit brought in on the soles of feet and must be kept clean to avoid significant wear. Above the floor, most objects are dusted and polished for mainly aesthetic reasons, and the greatest threat they face is often from the cleaning regime itself.

### Cleaning floors

Stone flags and memorials, marble and tiles are all vulnerable to wear. Start by minimising the risk of entry by clearing the area outside all entrance points of grit and dirt; install mats inside the doors for people to wipe their feet, and make sure that these mats are shaken regularly – away from the entrance points! Vacuum on low power rather than sweep up the dirt, and if cleaning seems unavoidable, consider how the water is likely to affect the material before you start to wash and scrub it! Sloshing large quantities of water around a stone floor can bring salts up from within the stone or from the mortar between them, forming ugly white blooms and may ultimately lead to damage. Marble may be ruined by washing with water.

**Tips for cleaning floors**

- Remove black marks left by rubber heels with a mixture of 300ml white spirit, 300ml water and a teaspoonful of Fairy Liquid
- Remove chewing gum by applying a block of ice to it until it becomes so cold that it can be chipped off
- When moving heavy objects, consider the construction of the floor and the effect of the load on the edge of stone slabs and tiles; spread the load across the floor with boards
- Make sure that solid floors can breathe across their entire surface so that damp does not accumulate: do not cover solid floors with impermeable materials or seal them with impermeable coatings
- Do not use wax on marble as it will ultimately yellow; dry polish marble floors with a polisher, using clean felt pads; brush marble and alabaster sculptures with a soft brush
- For cleaning timber, a cloth impregnated with paraffin and vinegar both refreshes the polish and collects the dust effectively

Adapted from *The National Trust Manual of Housekeeping* – see Recommended Reading

*Paths need to be swept around the entrance to a church to reduce the amount of grit being carried in*

### Dusting
Dust needs to be removed and not simply moved around the building, so it is best carried out with either a new, clean duster or a soft bristle brush for removing dust from crevices in one hand and the nozzle of the vacuum cleaner in the other. Take care to avoid damaging fragile material, and watch for splinters that snag, damaging the furniture. Old dusters are more liable to catch than new ones, and should not be used. If any damage or deterioration is noticed, make sure that it has been reported, and be very careful not to cause more damage if dusting it.

### Wax polishing
Do not wax unless you are certain that the surface has been previously waxed, and that the wax is not harming it. Even then, use wax sparingly and as rarely as possible: it does not 'feed wood' as some manufacturers claim. Never use silicone sprays: not only do they change the appearance of the polished surface, but they cannot be removed without first stripping and then resurfacing. If veneers or other details have lifted, avoid getting polish beneath them since if glue is to be used to repair the damage, the wax may prevent the glue adhering to its surface.

### Brass polish
Metal polish and proprietary brass cleaners contain abrasives and chemicals including ammonia in particular which damage the surface of brass, removing the patina. Brasses should be dusted or swept with a soft brush to keep them clean, not polished. If necessary they may be wiped with a rag soaked in white spirit to remove dirt and some stains.

### Cleaning windows
The inner surfaces of stained glass are usually painted. This surface is both precious and highly vulnerable to damage, and it should not be touched. Cleaning must be left to specialists. Leaded lights are generally vulnerable

The vulnerability of historic buildings and materials to damage by unconsidered actions should never be underestimated. With care and proper maintenance, buildings which have lasted hundreds of year may be expected to last for hundreds more, so we too may be able to hand over the properties entrusted to us by previous generations to generations to come.

Developed in Switzerland, **Classidur** paint is a topcoat and undercoat in one. 99% of the time **Classidur** needs no preparation and can be applied directly over most stains and surfaces* including whitewash, soot, fire damage, exhaust fumes and nicotine.

**Classidur** provides a totally white matt or satin finish and can be tinted with universal stainers if required.

Save time and money on your next job by using the best, **Classidur** – the class act.

*Non-cohering material must be removed and small section should always be test-coated first.

**BLACKFRIAR**®

Blackfriar Road, Nailsea, Bristol BS48 4DJ

**Tel: 01275 854911**
**Fax: 01275 858108**
www.blackfriar.co.uk

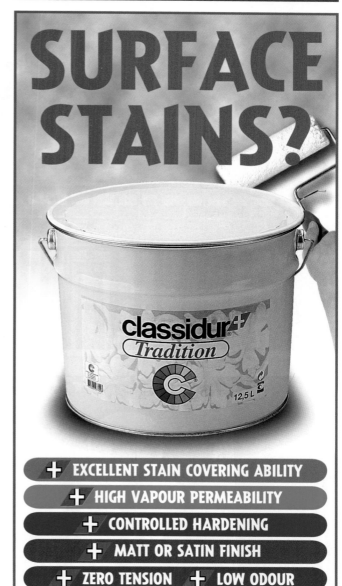

*Stained glass, carved marble and alabaster, ornamental ironwork and painted decoration – all vulnerable to inappropriate cleaning and all irreplaceable*

*Stained glass is particularly vulnerable and should only be cleaned by a conservator*

## MAINTENANCE CYCLES

Maintenance may be considered under three categories:

1. Routine, day-to-day maintenance and cleaning regimes to prevent deterioration and to repair any minor accidental damage or defects as they occur
2. Cyclical maintenance and repair works carried out over a longer cycle, such as cleaning gutters of leaves each autumn, or painting exterior woodwork every seven years
3. Occasional repair work, usually carried out strategically as part of a larger programme of repairs, such as repointing and roof repairs.

## RECOMMENDED READING


Mills, Edward D (Ed),
*Building Maintenance and Preservation: A guide for design and management.* 2nd Edition, Butterworth Heinemann, Oxford 1994 ISBN 0 7506 900 1

Sandwith H and Stainton S,
*The National Trust Manual of Housekeeping.* Penguin Books, London 1993


A variety of booklets and a video on maintenance issues are available from The Council for the Care of Churches, and more information is published on the Council's excellent new website, **www.churchcare.co.uk** – see *A-Z of Maintenance.*

## MAINTAIN OUR HERITAGE

Maintain our Heritage is a new organisation set up to press for a long-term, sustainable strategy for the care of our historic buildings, with the greatest priority given to maintenance. Maintain is to:

- mount a multi-faceted campaign to promote the wider understanding and adoption of maintenance. It is to seek to influence directly those responsible for historic buildings; and it is to seek changes in policy to encourage maintenance.
- pilot a not-for-profit service offering:
- an inspection of the fabric of the building;
- a maintenance action plan containing prioritised recommendations, in layman's terms, for maintenance and repair work;
- an explanation of the report in person; and
- the carrying out of a limited amount of first-aid on-the-spot repair where small, but critical, areas of disrepair are encountered during the inspection.

The first pilot is to be in Bath & North East Somerset in 2002.

For further information see
**www.maintainourheritage.co.uk**
or contact
Timothy Cantell, Project Co-ordinator,
Maintain our Heritage, Weymouth House,
Beechen Cliff Road, Bath BA2 4QS
Tel 01225 482074 Fax 0870 137 3805
E-mail tcantell@weyhouse.swinternet.co.uk

*St Anne's, Limehouse (Julian Harrap Architects)*

# STRUCTURAL ROOF REPAIRS

David Yeomans

THE STRUCTURAL TIMBERS of open church roofs can be breathtaking, whether they are highly decorated or simple, functional designs. Until the 17th century when it became common to introduce ceilings, the whole roof structure was most often designed to be seen from the floor below, and this form of roof was widely adopted by the Victorians in the churches of the Gothic Revival and in the numerous Victorian replacements for medieval originals. Open roof forms were also developed by 19th century architects such as Butterfield and Street to produce some highly original and imaginative structures.

The importance of these open roofs is obvious, but the historical and architectural importance of 'closed roofs' in which a ceiling hides all or most of its structural support, should not be overlooked. Although hidden above decorative ceilings, their roof structures are also considered to be of great historic importance.

These various roof types belong to quite different carpentry traditions using different materials, so they present quite different challenges in both inspection and repair.

## MATERIALS

Oak was the timber traditionally used for structural purposes in England until it became in short supply during the 17th century. From then on imported softwoods – that is to say, timbers from coniferous trees – were increasingly used. At first these came from the Baltic countries and then, towards the end of the 18th century, from North America. They are less durable than oak but they were easier to work and could be obtained in long lengths.

Wrought iron straps were commonly incorporated into the roof trusses of 18th century roofs and by the 19th century the use of iron in combination with timber was increasing. Roofs were then designed which relied extensively on iron fasteners and iron tension rods as well as the timber, even in those roofs that imitated medieval forms of carpentry.

## STRUCTURAL PERFORMANCE OF TIMBERS

The strength of a timber depends upon the species and upon growth characteristics. Strength is generally related to the density of the timber, so oak is stronger than the lighter softwoods, but it is also affected by the rate of growth, the presence of knots and the characteristics of the grain. For oak and other hardwoods the faster grown timber is stronger, so warm climates produce the best hardwoods, but this is hardly a relevant factor when we are dealing with native British trees. The best timber will have straight grain and be free from knots, and the small section oak timbers once obtained from coppiced woodlands had these properties. However, they also had a large proportion of the less durable sapwood, the lighter coloured wood closest to the bark. It was less easy to obtain large timbers of the same knot-free quality and early carpenters sometimes had to make do with pieces that had large knots. These can be a source of trouble.

Today we may be cutting small section timbers from much larger logs and while this will have the advantage that it will contain the more durable heartwood, the grain may not be so straight and it may contain knots. Timbers now have to be graded for strength by taking these factors into account.

In contrast to hardwoods, it is the more slowly grown softwoods that have the greater strength – the worse the climate the stronger the wood because there will be more growth rings for a given timber size.

Softwoods are similarly graded for strength but the difficulty here is in obtaining large-section timber. During the 18th and 19th centuries imported softwoods came from first growth forests. Today's managed forests are aimed at the wider demands of the construction industry where small section timber is required in quantity. If large section timbers are required for conservation purposes it may not be possible to match the precise species of the original and alternatives may have to be used. Sometimes second-hand timbers can be obtained from, for example, the demolition of 19th century warehouses.

One should note that when buying timber today of a particular grade, that grade refers to the worst part of the whole piece. One advantage of conservation work is that small quantities are normally being purchased for a specific location so that the strength required at that location can be assessed.

Perhaps as important to the conservator as the strength of timber is its moisture movement. In the past, oak was worked green with the effect that drying shrinkage took place after construction. Even today, because of the difficulty of drying large timbers, these too will commonly be worked green and the carpenter will be making allowance for shrinkage in the design of the repairs. Fasteners may be used as much to control moisture movement as to provide the necessary structural connection between timbers. However, movement may affect the appearance of repairs, and warping or checking (the development of splits in the surface of the timber) may also be considered to be a problem, even when the defect has no effect on its structural behaviour.

*St Nicholas, Denston, Suffolk: several of the 15th century timber beams visible here have been neatly repaired at the junction with the wall plate using new timber scarf-jointed to the existing, so that both new and old timber continues to function as an integrated structure.*

## UNDERSTANDING THE STRUCTURE

The roof covering and its drainage form the first line of defence in the preservation of the roof structure and although not the subject of this article, it is important to note that their proper design and maintenance are of paramount importance. Changes in the roof covering can affect the load on the structure.

To explain the structural behaviour of a roof requires an understanding of both its construction and the condition of the timbers; a knowledge of the history of structural carpentry will help here. Of course, the behaviour of a structure may change with time, either because of movements in the supporting structure or with movement and other changes in the timber. This means that it is not always possible to determine the way in which the structure is working, and the engineer may well have to consider more than one possible mechanism of load transfer when assessing how the structure is working now.

When there are signs of structural distress, the first step is to understand its causes and the structural mechanisms involved. In doing this one must recognise that the roof is part of the overall structure of the building and its behaviour may be related to the condition of supporting walls and their foundations. Sometimes this means that an examination of a problem which first appears in the roof may have to be extended to include a more general investigation of the building's structure.

The inspection and repair of open roofs is made much more difficult by the typical span of the roof and its height; often access to the timbers can only be gained with either scaffolding or a cherry picker. As detailed a knowledge of the structure as possible helps to avoid any unpleasant surprises when work begins.

## APPROACHES TO REPAIR

These early structures were not designed to conform to present-day structural codes and indeed may not do so. However, they have stood the test of time and the general principle for repair should not be to attempt to bring a structure into conformity with present-day codes. To do so may destroy historic fabric. Instead we should simply be giving the structure a 'helping hand' where necessary. The test of any intervention should not be whether it brings the structure 'up to date' but whether it is necessary. To avoid all unnecessary alterations, only the minimum should be done to ensure security. Of course, given the difficulty of surveying some structures, no matter how carefully the investigation of its condition the full picture may not emerge until work has begun. In such cases the repair plan may have to be revised accordingly.

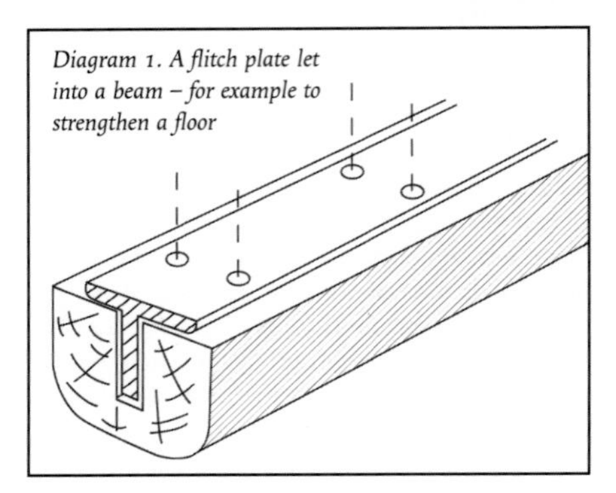

*Diagram 1. A flitch plate let into a beam – for example to strengthen a floor*

Where a roof structure is open to view from below, it is obvious that the successful solution must take into account the visual appearance of the repair itself. In contrast, in roofs that are hidden above ceilings, supplementary steel structures can be used without loss of visual quality and possibly with greater retention of historic fabric. In recent years a growing number of people recognise that the history of carpentry is of interest in its own right and that even some hidden structures may have a significance that needs to be respected in designing repair work.

The different approaches to the repair of timber structures are:

### Do nothing

This is presumably the most desirable course of action both for the survival of historic fabric and the cost to the client. Although the desire to be seen to do something may be a strong one, it should be resisted. Even where there is evidence of structural movement it may be long-standing and where such movement has ceased there may be no need for intervention. Where there is continuing movement it may not be immediately possible to assess the ultimate effect of this. Therefore a wait and see policy, with further monitoring, may still be more appropriate than taking immediate action, and the client should be willing to discuss such an approach with the engineer. Of course there are also those situations where some immediate temporary support may be required. *(Note that the filling of surface cracks in timber is a useless activity and should not be carried out unless there is some strong aesthetic reason for doing so.)*

### Strengthen the existing structure

Designing repairs that are not visually intrusive in open roofs presents a problem, as any cutting out of existing material obviously involves a loss of historic fabric and is to be avoided. Rather than designing carpentry repairs to replace decayed structure, it may be preferable to introduce some form of reinforcement. This can take the form of steel flitches let into beams. Steel reinforcements are commonly used to strengthen floors *(Diagram 1)* and in a roof this method has the advantage that it is invisible from below and can often be installed entirely from above. Steel reinforcing can be fixed either simply with mechanical fasteners or with the use of epoxy resins to supplement a mechanical fixing.

### Cut out and replace decayed or damaged timber

This may be done using carpentry techniques or by replacing lost material with epoxy resins. A variation on this approach is to substitute some form of steel structure to perform the function of the material that has been removed. The first of these methods may rely entirely upon traditional carpentry methods, but more commonly will use modern mechanical fasteners. These have the advantage that repairs can be designed with much less loss of original material while providing visually acceptable repairs. There is sill some debate about the suitability of epoxy resins for the repair of historic timbers but they are invaluable where there would otherwise be a loss of decorative surfaces. The replacement of timber with steelwork *(as in this heel joint – Diagram 2)* can involve the least loss of existing fabric but this approach is clearly only suitable where the repair cannot be seen, as in structures concealed above a ceiling.

### Introduce a supplementary steel structure

This has been done in roofs where the existing structure is of some historic importance. For example a steel structure was introduced by

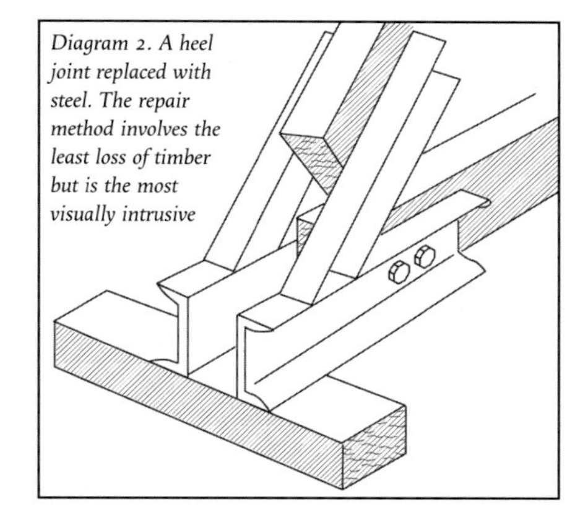

*Diagram 2. A heel joint replaced with steel. The repair method involves the least loss of timber but is the most visually intrusive*

Julian Harrap Architects into the roof of St Anne's, Limehouse *(Figure 1)*.

The basic principle for historically important structures of preserving as much of the historic fabric as possible may conflict with both the need to preserve the appearance of the structure and the requirements of structural performance. Finding a suitable compromise between these conflicting requirements is often a matter of judgement, and the most successful solutions are often the result of a close collaboration between engineer and carpenter. The carpenter will always be able to advise on the practicability of a repair and may even take the lead in suggesting a possible design. The need for such a close working relationship should be borne in mind in the selection of the conservation team.

### THE IMPORTANCE OF THE TEAM

The design of even modern timber structures is a comparatively specialist activity and this is even more true of timber repair work. It will be clear from what has been said above that one should be looking for engineers who have some knowledge of historic structures and if possible an awareness of their historic significance. Similarly one should be looking for contractors with appropriate skills. Not all will be used to using modern timber fasteners and the bad reputation that epoxy resin methods once had was a result of the poor repairs carried out by inexperienced contractors.

It has also been suggested that there should be a close collaboration between the parties involved, including the contractor, even at the design stage. This implies contractual arrangements that allow such collaboration rather than the normal competitive tendering. Also, where advice to do nothing might be the desirable outcome of a structural investigation and analysis, payment of consultants on a time charge basis, rather than as a percentage of the cost of works, is clearly necessary.


**DAVID YEOMANS** is an historian and an engineer specialising in the repair of historic buildings. He is Chairman of the ICOMOS UK Wood Committee which will shortly be producing guidelines on the repair of historic timbers. E-mail david@yeomans.u-net.com

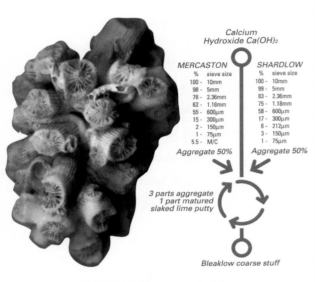

*The repaired lych-gate*

# THE HESTON LYCH-GATE

## A model timber frame repair

ACCORDING to a plaque fixed to it, the lych-gate at St Leonard's church, Heston, near Hounslow in Middlesex dates from *c* 1450. It is a particularly magnificent oak-framed structure, with a tiled roof supported by two heavy braced posts. But what makes it stand out from other examples of its time is its centrally pivoted gate and its rather quirky self-closing mechanism.

Although only a small building – if it can be called that – its conservation posed challenges typical of a much larger project. During the 1970s it was decided that a number of the key principal members required replacement. This was achieved by avoiding the re-use of a number of mortise and tenon connections and instead relying on skew timber dowel rods and bolts. This method proved disastrous in the long term with the structure open to the prevailing weather conditions. The recent joint connections had deteriorated affecting the stability of the structure and ultimately fracturing some of the principal structural members.

In 1998, McCurdy & Co Ltd was approached by the parochial church council of St Leonard's to advise on the repair of the structure, 'to reinstate the long term structural integrity of the building'. A 25-year lifespan for repairs was not acceptable.

### HISTORY

'Lych' is the Anglo-Saxon word for corpse and, according to tradition, lych-gates were designed to give shelter to the pallbearers at the entrance to the graveyard. The Heston lych-gate is a well-known example, partly for the design of its pivoted gate which was described by JL Andre in 1896 as a rare example of a *Tapsel* gate. This was designed to close itself by means of a counterbalance discreetly located in the roof above. As the gate was opened, the central pivot rotated, drawing a chain or metal strap which wound around a wheel at the top of the pivot post. This chain ran around a wooden pulley wheel next to it, up almost to the apex of the roof, over a second wooden pulley wheel and then down to a weight suspended in mid air. The weight rose as the gate opened, and when the gate was released the weight descended again, drawing it shut.

By the time the current program of repairs was commenced the counter balance had been lost. However, all the other elements had otherwise survived, and although a vehicle had damaged the gate shortly beforehand, it was clear from these components how the gate had operated. An elderly local resident whose father was involved in maintenance around the churchyard remembered there being a stone and described how a metal strap around its circumference was then attached to a chain. This account was further corroborated by JL Andre's 1896 description: "At Heston in Middlesex, where the Tapsel gate is an elaborate one, placed beneath a lych-gate, and made to shut by means of a large wheel, round which passes a large chain, with a lump of stone at the end acting as a balance weight, the whole arrangement forming a very picturesque and quaint object".

## THE LYCH-GATE BEFORE WORK STARTED

### Typical problems

*Traditional mortise and tenon joint substituted for metal strap, which had failed*

*Loads from wall plate now carried by only the bottom half of the tie*

*Prop introduced to support broken tie beam due to failure of metal strap*

*Traditional mortise and tenon joint substituted for timber dowels, which had failed*

*DPM introduced in 1974 causing cill plate to decay by catching rainwater*

*Above: The lower pulley wheel and the pivot strap*
*Left: The gate after it had been damaged by a car*

It is possible that the lych-gate is not as old as its plaque states, but McCurdy & Co who investigated its history before commencing work, were unable to find any conclusive evidence for or against the date given, 1450. English Heritage instigated the use of dendrochronology as a means of dating the building but unfortunately a suitable sample could not be obtained.

The earliest reference to the lych-gate found by McCurdy & Co was a lithographic print of 1862 which had been drawn up as one of a series of scale drawings for an architectural dictionary. This shows the lych-gate in its present form shortly before the main Victorian alterations to the church were carried out in 1865.

Like many medieval churches, St Leonard's was fairly extensively remodelled by the Victorians to conform to their idea of how a medieval church should look. According to survey notes made by the National Monuments Record (NMR) in 1937, the church tower is all that remains of the original medieval church structure. There is every indication that the west porch, which overlooks the lych-gate, was reconstructed using original materials at this time, but there is no indication that the lych-gate was reconstructed at the same time.

## PAST ALTERATIONS AND REPAIRS

From photographs taken by the NMR in 1937 it can be seen that the lych-gate had undergone a repair phase prior to those undertaken in 1974, and it is possible that some of these were carried out in 1865. Scarf repairs were clearly visible on one of the braces, and one of the tie beams was obviously a replacement. Other repairs included the wall plates and the rafters, and the pivot gate was shown as having additional straps to the corner joints and a partial replacement of one of the stiles.

In 1974 when the gate was repaired again, the vertical boards shown on the gate in earlier photographs were removed and not reinstated. It is also clear that the braces and the stiles were replaced as, unlike the top rail and central post, they do not have a groove for the vertical boarding.

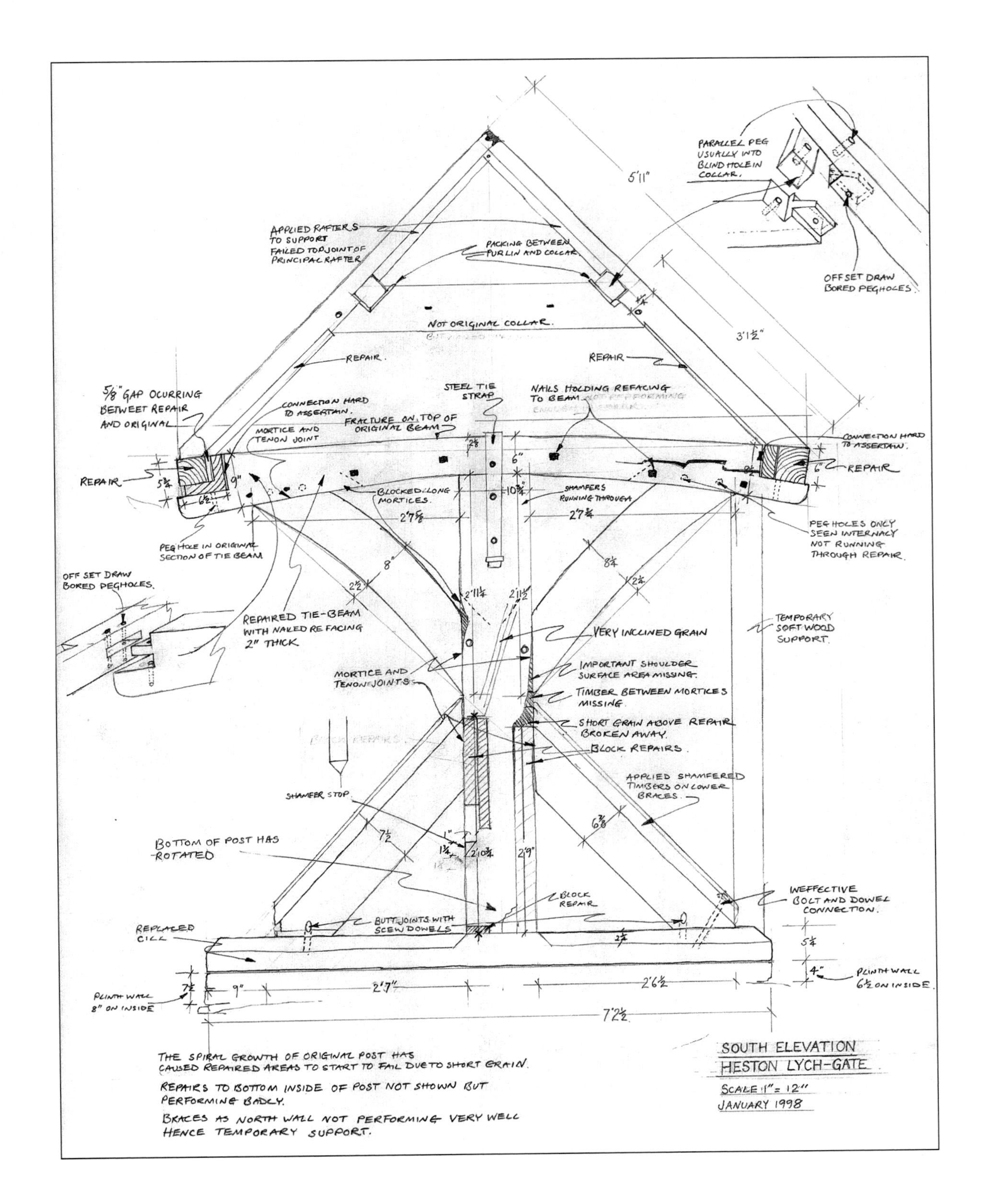


The repairs carried out at this time also included replacing both cill beams of the lych-gate and the north post. The jointing methods used for the north post suggests that this work was carried out *in situ* as the braces, which were originally tenoned top and bottom, now have a tenon at one end only, with the other end butted up against the new post and secured with dowel rods. The new method of connection was a poor compromise which did not stand up to the rigours of movement and exposure to weather, and gradually the connections were becoming loose and ineffective.

Many of the repairs carried out at this time relied on nails and glue, with inevitable consequences. Externally glue is liable to break down readily or trap moisture causing the timber to decay. Block repairs which had been inserted to repair the ends of the two wall plates did not incorporate the original mortise and tenon jointing detail, relying instead on metal straps to tie the wall plates back to the tie beams. In one instance the metal strap had corroded through, making the connection ineffective. Subsequently some of the timber blocking pieces had also become loose. The inappropriate choice of fixings and glue, together with the design of the repairs, had affected the structural performance of the wall plates.

The tie beam on the south elevation had been extensively repaired with a two inch thick, full length, refacing. This has been inadequately nailed to the remaining original beam and subsequently it had split along the bottom line of the wall plate housing. This split had occurred primarily because the brace below had failed to transfer the load of the roof down into the post. The brace had dropped because the bearing surface in the post mortise had become soft and rotten. Consequently a

*Detail showing the broken tie beam supported by the prop. The broken metal strap which caused the problem runs along the top of the tie beam, so it cannot be seen*

*South post viewed from the west*

softwood prop had recently been inserted to provide the necessary support.

## SCOPE OF WORK

By the time work commenced on site in the winter of 2000, much of the work carried out in 1974 was showing signs of progressive deterioration, with joints pulling apart causing the frame to move unduly, and the lych-gate was rapidly becoming unstable, as the temporary softwood prop clearly indicated.

McCurdy & Co advocated a return to traditional mortise and tenon joints for new members, retaining as much of the existing as possible, including repairs. For retained members they advocated slip tenon or bare faced tenon repairs (see diagram) where required. Where straps and bolts would still be required to restrain joints these were to be replaced using more durable modern materials.

To carry out such a thorough program of repairs it was decided that the most efficient manner would be to dismantle the timber frame. So, before any work commenced, the existing structure was carefully recorded by measured drawings and photographs. The structure was then dismantled, setting aside the existing clay roof tiles for reuse, and the timbers were transported to the workshop for further examination, before determining which elements could be repaired and which would have to be renewed.

New timbers were required for the cill plates, north post and a number of braces. Freshly felled green oak was selected to replicate the original timber sizes and conversion from the log. Boxed heart for example was the conversion used for the north post. Replacement curved members were selected from naturally curved swept boards with similar radius of grain. On these new timbers, the original joints were reinstated and secured with tapered cleft oak pegs, as they would have been originally. Features such as chamfers and gate stops, which are still evident on the surviving original members, were also reinstated on replaced timbers.

Existing timbers were retained, whether or not they were original. Repairs to these timbers were made using air-dried oak to match the grain and moisture content of the existing oak, therefore minimising differential movement. Freshly felled oak was used for the new framing. As this timber had been milled by machine, any mechanical marks that were apparent were removed by hand plane. This surface treatment follows that originally used on the lower primary pit sawn timbers whose surfaces have been dressed back. It also provides a similar surface patina to that naturally left on the hewn boxed heart timbers.

When the gate was struck by the vehicle, most of the shock was absorbed by the failure of the earlier repairs which had already worked loose, so it was possible to salvage almost all of the original material. The decision was taken not to restore components and features unless the component had to be replaced for structural reasons, so the vertical boarding was not reinstated, nor was the bracing replaced with a component to match the original more closely. Repairs were therefore limited, largely confined to remaking previous repairs. By extending the length of some of these repairs it was possible to restore integrity back into the individual members. The reconnecting of the original components again used traditional joint methods where possible.

The repaired timber frame was bedded onto the retained stone plinth wall using porous lime mortar. The impervious plastic membrane used previously had contributed to premature onset of decay by trapping rainwater.

The lych-gate has been re-roofed re-using the removed clay tiles and where required supplemented with reclaimed tiles to match these have been laid on cleft oak laths as would be expected on a building of this period.

When finished at Easter 2001, the structure retained as much of the original fabric as possible without compromising the structure's long-term viability. All the alterations to the existing structure were based on a thorough investigation of its history and a clear understanding of what had been replaced and when. Although the new components replicated as closely as possible the original details such as chamfered stops, decisions to replace components were made on structural grounds alone, and not for the sake of restoration. Only by adopting this sort of approach to the repair of historic buildings and structures can we hope to pass on to future generations buildings which are truly historic.

This article was prepared with the help of **JULIAN LADBROKE** of McCurdy & Co Ltd

Bath Abbey

# LISTED BUILDINGS

## A Quick Guide to the Statutory Protection of Historic Places of Worship

*Like many memorials in churches and chapels throughout the country, this simple, but superbly executed example at Bath Abbey provides a snapshot from the history of one family and of the British Empire*

CHURCHES account for 11 per cent of all listed buildings in England, but of the most important (Grade I) listed buildings, churches account for 40 per cent. That such a high proportion of the most important historic buildings in this part of the UK is accounted for by churches should not seem astonishing when you consider the number of country towns and villages with medieval churches, and the quality of church architecture and ornamentation. From Norman fretwork to Modernist minimalism, church architecture of every age has always attracted great designers and the very best craftsmen.

As the Church of England's excellent website puts it:

> *These buildings are first and foremost places of worship, witnesses in stone and brick to the truth of the Gospel. But these buildings are much else besides: their spires and towers, rising over town and countryside, play an important part in defining the English sense of place, community and identity. From modest church to glorious cathedral, they are a priceless part of our national heritage. England would be markedly different without its much loved churches and cathedrals.*
>
> www.england.anglican.org

### THE PROTECTION

All buildings in the United Kingdom are protected to some degree by the requirement for planning permission. This controls the development of new buildings and the most significant alterations to the exterior of existing ones, but it does not provide effective control over demolition or alterations to the interior. Therefore historic buildings are further protected by being 'listed' in their own right or included within 'conservation areas'. Both these forms of protection cover demolition, and all alterations are controlled if the building is listed.

> *No person shall execute or cause to be executed any works for the demolition of a listed building or for its alteration or extension in any manner which would affect its character as a building of special architectural or historic interest, unless the works are authorised.*
>
> Planning (Listed Buildings and Conservation Areas) Act 1990, Section 7, and – Planning (Listed Buildings and Conservation Areas) (Scotland) Act 1997, Section 6

A listed building is one which is included on the statutory list of buildings of 'special architectural or historic interest' which is maintained by the Government. This means that almost all alterations to the building, its boundary walls and other structures within its grounds can only be carried out with special permission. For altering listed secular buildings and some listed ecclesiastical buildings, the permission required is 'listed building consent'.

Conservation areas are designated by the local authority. The principal effect of the designation is that conservation area consent is required for the demolition of a building within a conservation area. In addition, applications for planning permission are affected and the policies of the local authority should be carefully noted as local authorities are required to pay special attention to 'the desirability of preserving or enhancing the character or appearance of that area' when considering an application for planning permission.

In addition to these protection measures, some features within the grounds of an historic church may also be scheduled as 'ancient monuments'. These are usually features which cannot be described as buildings, such as an ancient burial mound or a standing stone. All works affecting a scheduled monument or the ground surrounding it require 'scheduled monument consent'.

Conservation is enforced by prosecution as it is a criminal offence to carry out work to

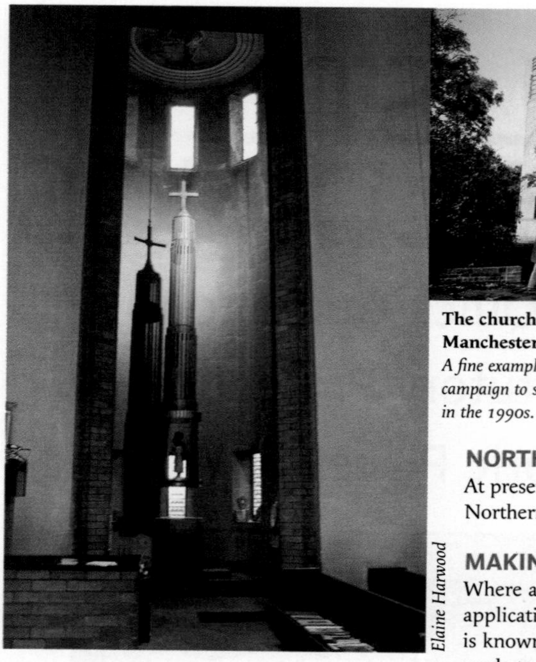

Elaine Harwood

a scheduled monument, a listed building or a building in a conservation area without the permission required.

## ECCLESIASTICAL EXEMPTION IN ENGLAND AND WALES

Prior to 1994 all churches and chapels in England and Wales were exempt from listed building and conservation area controls. Since then, 'ecclesiastical exemption' as it is known, has been restricted to churches and chapels of the six denominations operating an acceptable internal system of control, provided that the building remains in use as a place of worship. These are: the Church of England, the Church in Wales, the Methodists, the Roman Catholics, the United Reformed and those Baptist churches where the Baptist Union acts in the capacity of trustee.

The exemption does not effect the need for planning permission, which is required for most alterations to the exterior of a building, whether it is listed or not.

As the exemption only relates to buildings which are *in use*, proposals to demolish a place of worship also require the usual consents, since by definition the building cannot be in use at the time it is demolished.

## SCOTLAND

Ecclesiastical exemption in Scotland, which was limited to listed building consent only, has also now been reassessed and a pilot scheme was introduced on 1 January 1999 for a trial period of three years. Under this scheme a congregation wishing to carry out works to the exterior of a listed church building is required to submit details of the proposal to the planning authority and Historic Scotland and to obtain their agreement before undertaking the work.

For the moment, works to the interior of a listed church building remain exempt.

Full details of the pilot scheme may be obtained from the relevant planning authority and Historic Scotland. If the scheme is successful, it (or a suitable modification of it) will be made permanent.

Elaine Harwood

**The church of St Nicholas in Burnage, Greater Manchester, by Cachemaille Day, 1930-32.**
*A fine example of a 20th century listed church. A popular campaign to save the church successfully averted its demolition in the 1990s.*

## NORTHERN IRELAND

At present alterations to all churches in Northern Ireland retain ecclesiastical exemption.

## MAKING AN APPLICATION

Where a place of worship is listed, an application for consent (or a faculty as it is known within the Church of England) is made to either the local planning authority (if the church is not exempt) or the Diocesan Advisory Council or the Cathedrals Fabric Commission (if it is Anglican), or the equivalent board (if it is another church that is exempt). All the organisations which would usually have a statutory role in a listed building consent application are consulted, including the statutory authorities (English Heritage, Cadw, Historic Scotland or the DoE Northern Ireland) and the amenity societies (Ancient Monuments Society for example). Generally, if the proposal does not affect the character of the place of worship as a listed building, or if the affect can be justified on conservation grounds, the alteration may be approved, but for many applications there follows a period of careful deliberation and negotiation to achieve a solution which satisfies the needs of the church as best can be, without permanently harming that which makes the building special.

## PRINCIPLES

The criteria used to assess a proposal are set out by the Government in 'guidance' and 'directions' (see *Recommended Reading* below). These are based on internationally recognised principles described in the documents and in BS 7913: *Guide to the Principles of the Conservation of Historic Buildings* published by the British Standards Institute. The following principles taken from the Australian *Burra Charter* which was last updated by ICOMOS (the International Council on Monuments and Sites) in 1999 highlight some of the key issues which lie at the heart of all these documents and are particularly relevant:

- *The policy for managing a place must be based on an understanding of its cultural significance ...and its physical condition.* [Article 6.1-6.2]
- *Existing fabric, use, associations and meanings should be adequately recorded before any changes are made to the place.* [Article 27.2]
- *Conservation is based on a respect for the existing fabric, use, associations and meanings. It requires a cautious approach of changing as much as necessary but as little as possible.* [Article 3.1]
- *Changes which reduce cultural significance should be reversible, and be reversed when circumstances permit.* [Article 15.2]
- *Contents, fixtures and objects which contribute to the cultural significance of a place should be retained at that place...* [Article 10]
- *Reconstruction is appropriate only where a place is incomplete through damage or alteration, and only where there is sufficient evidence to reproduce an earlier state of the fabric. ...Reconstruction should be identifiable on close inspection or through additional interpretation.* [Article 20]

These principles provide useful guidance on what might be acceptable. However, bear in mind that the last thing that the authorities want is to see a place of worship become redundant:

> *English Heritage wants to see churches maintained and kept in use, and recognises that churches in active use must adapt to the needs of the time. Our main concerns are to make sure that the historic fabric and interest are respected and that nothing of value is irretrievably lost as a result of new works.*
>
> www.english-heritage.org.uk/knowledge/conservation/churches.asp

Listing does not mean that a building cannot be altered or even demolished: indeed between 1968 and 1999 no less than 85 listed Anglican churches were demolished, 311 were taken into the care of the Churches Conservation Trust and 873 were appropriated for other uses. However, it does mean that all proposals to alter or demolish the building are carefully considered.

## RECOMMENDED READING

Mynors, Charles; *Listed Buildings and Conservation Areas*, Third edition, Sweet and Maxwell, 1999

*Guide to Historic Buildings Law*, Cambridgeshire County Council Planning Section, 1997

*The Ecclesiastical Exemption Order: What it is and how it works*, The Department of National Heritage and Cadw, 1995 (available free of charge)

*Scotland's Listed Buildings: A guide to their protection*, Historic Scotland, 1998

*Listed Historic Buildings of Northern Ireland: An owners' guide*, DoE Environment and Heritage Service, 1994

*The Burra Charter*, ICOMOS, Australia, 1999 (see www.icomos.org/burra_charter.html)

BS 7913: *Guide to the Principles of the Conservation of Historic Buildings*. The British Standards Institute, 1998

## GOVERNMENT GUIDANCE

(Full texts of UK legislation can now be found on the Internet at www.hmso.gov.uk)

**England**

Planning Policy Guidance Note 15: Planning and the Historic Environment

**Wales**

61/96 Planning and the Historic Environment: Historic Buildings and Conservation Areas

1/98 Planning and the Historic Environment: Directions by the Secretary of State for Wales

**Scotland**

Memorandum of Guidance on Listed Buildings and Conservation Areas

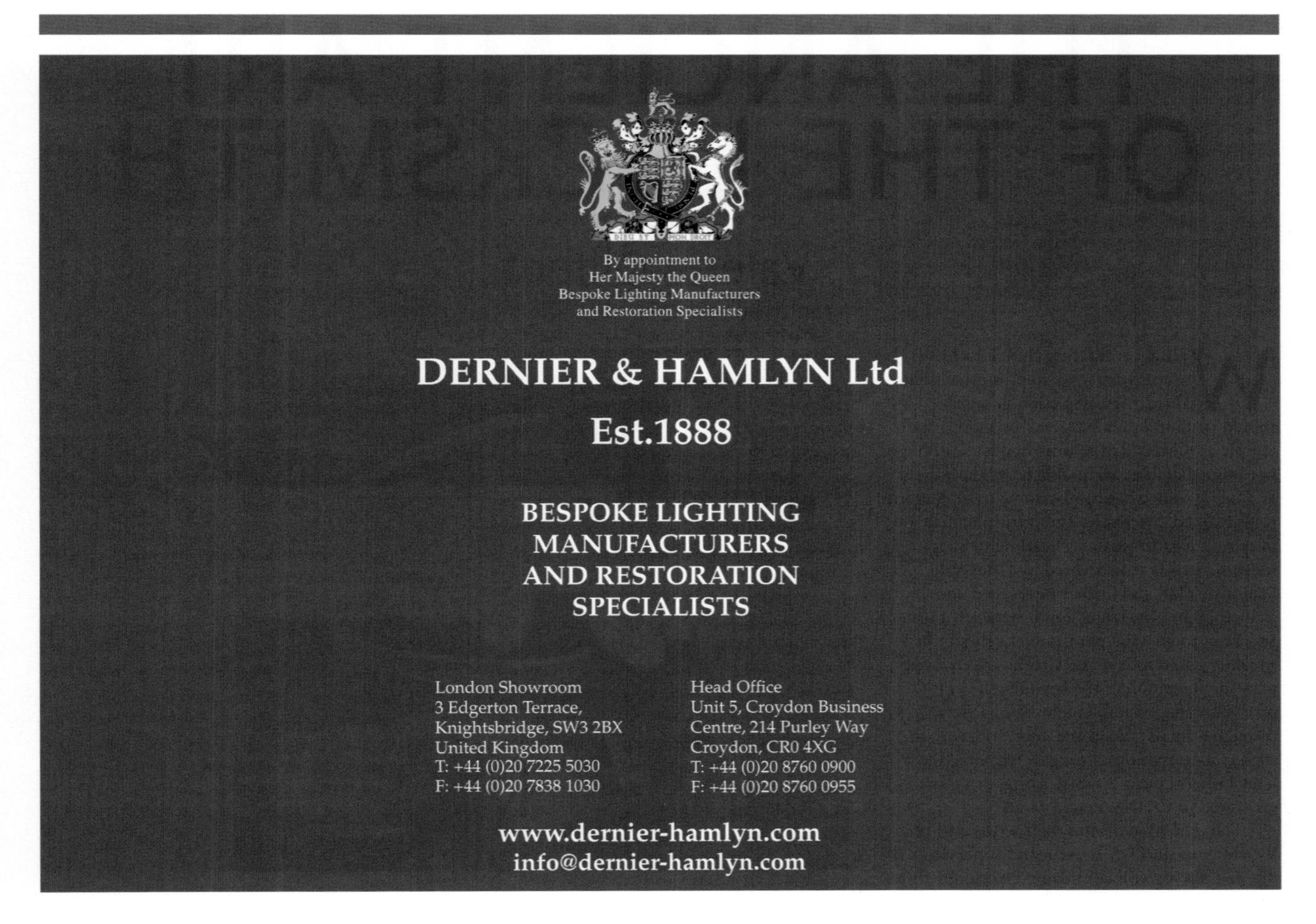

Tel: +44 (0)1403 78 5141
Fax: +44 (0)1403 78 4663
Email: sales@lambsterracotta.com
Website: www.lambsterracotta.com

# THE ANCIENT ART OF THE LOCKSMITH

## Valerie Olifent

WHEN considering church locks it seems particularly appropriate that the earliest depiction of a lock should be found on a bas-relief in an Egyptian temple at Karnak dating from 2000BC. Although not immediately recognisable to modern eyes, and being somewhat cumbersome in operation, it functioned effectively. The principle, that of raising pins to create a shearline to allow movement, was rediscovered by Linus Yale Senior in 1848 and further developed and refined by his son Linus Junior between 1861 and 1865 to give the pin tumbler cylinder lock so widely used today. The Greeks are credited with the invention of the keyhole, the point of a sickle shaped implement being inserted through a small hole in the door, and, with a slight rotary motion, closing or withdrawing a large bolt. A Linear B tablet dating to 1300BC, excavated in Crete, was translated

Thus the Mayors and their wives and the Vice-Mayors and key-bearers and supervisors of figs and hoeing will contribute bronze for ships and the points of arrows and spears.

Keys are mentioned in the Old Testament, notably in Judges ch3 v25, written around 1170BC and Isaiah ch22 v22, from about 740BC. The earliest lock excavated came from the Palace of Sargon at Khorsabad in Iraq, dating from 700BC. By the time that Vesuvius erupted in 79AD, when a metal worker's shop was overwhelmed, locks had been developed and had assumed a form recognisable to modern eyes. Many have been excavated both from Pompeii and from the numerous Roman sites in Europe and the Middle East. As they were now made from metal a large number have survived. Padlocks with a spring tine mechanism were found at York when the Jorvik Viking settlement of 850 was discovered.

A small but useful source of information from this period through to medieval times comes from the art of the period; carvings, wall paintings, illuminated manuscripts and stained glass. Depictions of everyday life sometimes show contemporary locks and keys and portrayals of St Peter can be a rich source. Even the Bayeux Tapestry shows Duke Conan of Brittany surrendering the keys of the town of Dinan to William on the point of his lance. Written records start to appear in the medieval period. The surviving accounts for the refurbishment of Portchester Castle in 1385 record the purchase of locks, and in 1394 London smiths were forbidden to make keys from an impression "by reason of the mischiefs which have happened". In 1411 Charles IV of Germany created the title of 'Master Locksmith' and by 1422 the London Guilds included the 'Lockyers'.

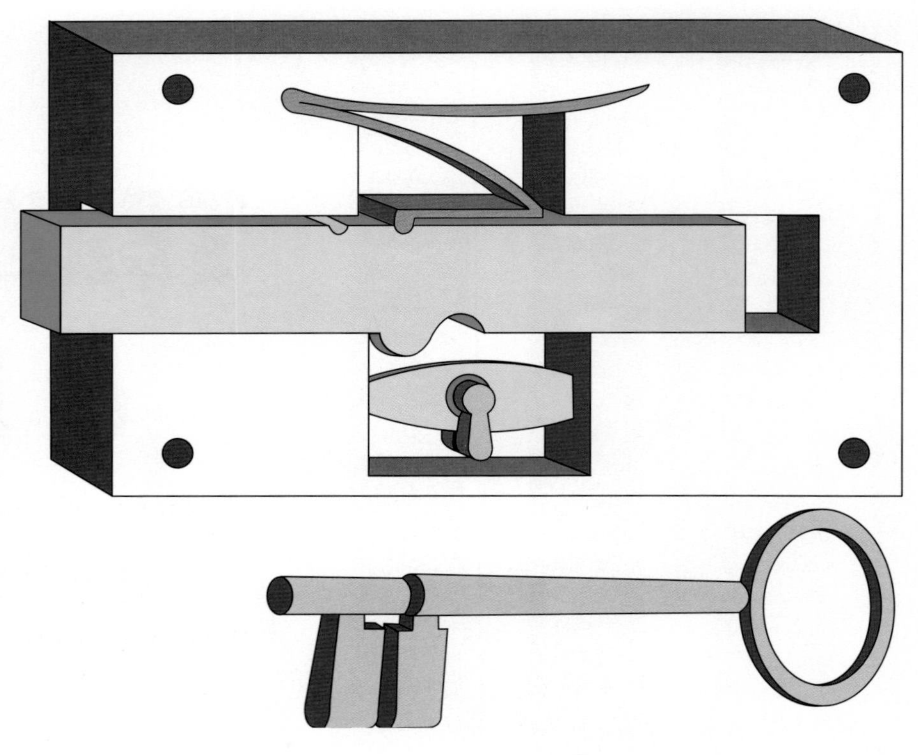

***Figure 1***

*THE BANBURY LOCK*

*Above, a close-up photograph of an 18th century Banbury lock mechanism (cap removed) from Theddlethorpe Church in Lincolnshire which is in the care of the Churches Conservation Trust. Note the decoration to the tumbler and spring.*

*Below, a diagram of the Banbury lock showing mechanism set into a wooden stock, and the distinctive key with a collar set within the width of the bit*

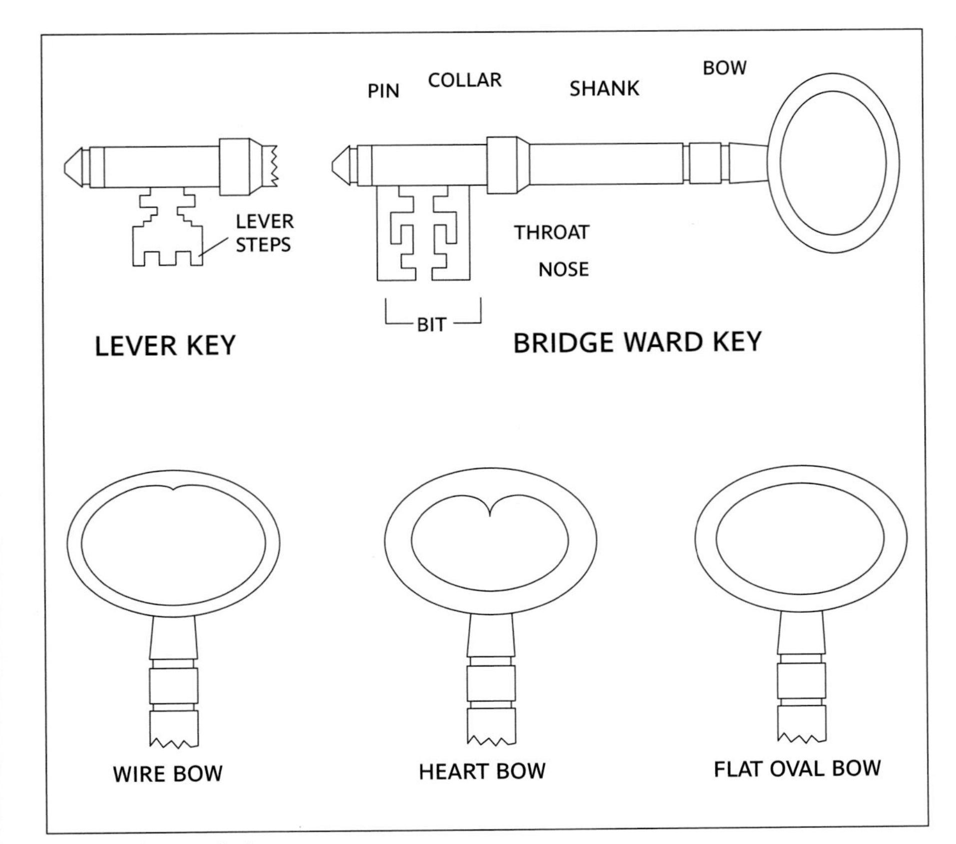


**Figure 2** *The parts of a key*

Some locks still in use do survive from this time, in historic college and university buildings as well as churches, but most are in private collections or museums. In the Victoria and Albert Museum you can see the 'Beddington Lock' which accompanied Henry VIII on his travels through the kingdom, being installed on his chamber door wherever he stayed to ensure his security and privacy. In the accounts for July 1532 is written *Item – paid to the smythe that carryeth the lock about wh the King in reward VIIsVIc.* After the distress of the Civil War and the privations of the Commonwealth, the Restoration of the monarchy in the 17th century saw a flowering of architecture and the arts, which extended even to locks and keys. Locks were made of an intricacy and beauty rarely equalled, often with the mechanism as highly decorated and engraved as the case.

Even until the mid 18th century and beyond, when elegance ruled, a degree of decoration of the mechanism sometimes persisted, enclosed within the plain, simple lockcase. The latter half of the 18th century saw the beginnings of the Industrial Revolution. In 1778 Robert Barron took out a patent to improve the security of locks, earning him the appellation 'Father of the English Lever System' *(see Figure 5)*. Some of the locks with his 'modern' and distinctive mechanism and keys can still be found in churches. In 1784 Joseph Bramah, the inventive Yorkshireman, patented the Bramah lock, an entirely new concept in lock design which utilized a series of sliders in a circular pattern to provide exceptional security. Building upon these major advances the 19th century saw a proliferation of patents for 'new and improved' mechanisms and developments, not all of which have stood the test of time. This period also saw advances in the manufacture of key blanks. Previously all hand forged, the advent of water and steam powered drop hammers brought a stamping process to key making, superseded by the discovery of malleableising cast iron which came into use for casting key blanks from around 1816: these processes are reflected in the changing shape of key bows as mass production was introduced.

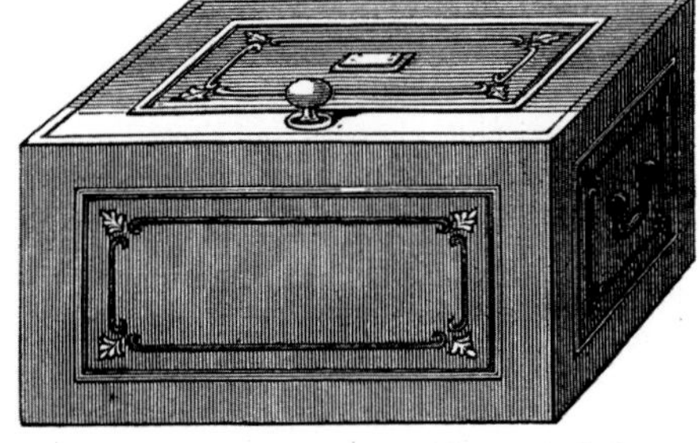

**Figure 3** *A cast iron safe: this safe is typical of those produced in response to the George Rose Act of 1812*

## CHURCH LOCKS

Many different types of locks can be found on church doors, depending on age, location and patronage. They may be either rim locks, mortice locks or even padlocks, and be of completely metal construction or of wood and iron. The doors of many historic churches still carry an old wooden lock although often you find that a modern 5-lever mortice lock has been installed along side it to meet insurance requirements. Some of these old locks will date from the foundation of the church and some from when the original door was last replaced, but many are the result of a Victorian makeover.

The generic name for the family of wooden locks is 'woodstock locks' a term dating back to the specialisation of trades springing from smithing, with some terms in common between the various branches. The earliest form is the 'Banbury lock' *(Figure 1)* in which the wooden stock is an integral part, with the metal components of the lock being mounted in the wood; the key is of a very distinctive form, with the collar being located within the width of the bit. Before industrialisation many Banbury locks were made by local craftsmen and so, especially amongst early ones, there are many variations to the standard deadbolt pattern, some having latchbolts in addition, operated by the key; some having double bolts and some double-handed for use in either a

**Figure 4** *Plate lock from the 18th century showing hand cut stock and dovetail plate*

**Figure 5** *Late 18th/early 19th century high security lock from a Butler's Pantry, dismantled to show complex warding and Barron's double tumblers. The keys are reproductions made to fit the lock.*

right-hand or left-hand application. There were also great differences, especially in the 18th century in the external shape of the keyhole, necessitating special keys, an additional security measure.

Improved iron working techniques in the 18th century were used in the 'plate lock', when all the working parts were mounted on a metal plate and the wood was merely a cover. Plate locks also had a long history of development, characterised by the shape of the metal plate which reflected the increasing degree of mechanisation in forming the corresponding recess in the stock. This progression from dovetail shape to rectangular, to rectangular with rounded corners, and to semi-circular end, clearly illustrates the stages of development from hand made artefacts to the products of industrial processes via the ever increasing use of powered cutters. However, as the industry as a whole, with a few notable exceptions, was composed of small 'backyard factories' there was a considerable degree of overlap between all these various features which can make dating very difficult; for instance a modified version of the Banbury lock was made up to the end of the 19th century and even later. Woodstock locks were not left out of the proliferation of lock patents, there being several incorporating greater security, easier operation and different methods of manufacturing the various components.

One further development was fuelled by the Victorian passion for refurbishing churches and for things 'gothic'. This was the 'church door lock', a specific item which incorporated a latchbolt and ring handle as well as the deadbolt. As they were designed for use in this burgeoning market they are usually of much heavier and better quality, with varnished oak stocks and often highly decorated with chamfering and ornamental metal work. There was even a small market for imitation medieval locks, with a large and splendid stock covering a very ordinary standard lock mechanism. Woodstock locks seem to have been the preferred lock for church doors until the 20th century and the need to meet modern security requirements. However, iron case rim locks, especially the better quality ones from the latter half of the 19th century were sometimes used on new build churches and also mortice locks for those with a 'Modern' style or new doors.

Locks were not used solely on the main doors; there will usually be locks on bell towers, vestries and organ lofts too, often as old as the church, sometimes remaining forgotten during a modernisation. Ornamental wrought iron screens and gates enclosing private chapels and mausoleums were often fitted with specially made locks decorated in matching style, while outer porch grille gates have stout locks, although, being exposed to the weather, few have survived in working order, often being replaced by a modern padlock. Then there is the parish chest which traditionally had three locks. These could be any combination of locks and/or padlocks, with the vicar and church wardens each having the key to a different one so that all three needed to be present to gain access. Some parish chests date back many centuries but not many of the locks remain. Some small churches also retain an ancient and crude alms or poor box, sometimes with provision for two locks, but these locks have rarely survived, having been superseded by modern padlocks or replaced by small 19th century and 20th century wall safe donation boxes. There are often small locks on aumbry doors (an aumbry is the small niche close to the altar in which the sacrament was kept), some of quite high security, and there is the ubiquitous cast iron safe installed in response to the George Rose Act of 1812 requiring church registers to be kept safely in an iron box. These safes are fitted with locks which have very complex warding *(the gate through which the key passes when it is turned – see Figure 5)*, requiring skilfully made keys of the highest standard, carrying on the tradition of Georgian strongroom keys.

## CONSERVATION, CARE AND MAINTENANCE

Locks are very often forgotten in any conservation and maintenance programmes, being regarded as totally utilitarian objects, while other aspects of church architecture and decoration of a similar age are looked upon as remarkable. Because, upon the whole, very little is known of their history and development and there is very little published material to guide those interested, most people do not realise what exceptional artefacts are in their care.

**Figure 6** *Gate lock to 1868/9 South Transept memorial chapel in St Peter's Church, Deene, Northants, in the care of the Churches Conservation Trust. The design of this lock echoes the style of the decorative ironwork. It was refurbished by The Keyhole and three reproduction keys were made for it.*

Because they are usually strongly made and can withstand an enormous degree of neglect no notice is taken of them until they cause trouble. It would perhaps be useful to include them in the 'quinquennial inspection', the essential survey of the fabric carried out every five years.

To ensure that problems are discovered in time, examine all the keys for signs of wear such as deep grooves on the nose of the key or on either side of the bit; these can lead to poor location of the key in the mechanism and difficulty in operation. Look at the keyholes, have they rusted or worn oval so that the key flops about, another cause of poor location. Remember always that a key and a lock work together and wear together, so when having a key duplicated if it is worn the result would be two worn keys! It is always better to make a key to the lock than to copy from a pattern key: locks can usually be fairly easily removed from doors, and, while the new key is being made, can be refurbished to bring them back to much easier and more efficient operation.

Test the lock by locking with the door closed, if resistance is felt, try again with the door open. If the bolt extends freely, check the alignment of the bolt with the staple on the doorframe; the door may have dropped due to wear on the hinges or loosening of the wood joints and the bolt may be fouling the bottom of the staple. The closing of the door may have been affected by swelling of the wood in wet weather or damp conditions, again, causing the bolt to foul the staple; if only a minor or temporary degree of distortion is present, learning to exert firm pressure on the door whilst turning the key is usually sufficient to overcome the problem. Still with the door open, operate the lock so that the bolt is extended; if the bolt can be pulled further out until a sharp click is heard, this indicates wear to either the key or the lock mechanism or both and that the bolt is not seating correctly, which will eventually lead to a lock-out situation; if the bolt can be moved freely to and fro, usually the spring is broken and as above could lead to a lock-out, but in addition this fault compromises the security of the lock. In both cases advice should be sought urgently.

The front limb of a large key will often be bent slightly in towards the centre cut of the bit, being the piece which normally hits the ground first when the key is dropped; this should not be straightened unless absolutely necessary, as it is liable to fracture. Check that it is not cracked as well; if it is going to break off it usually does so inside the lock during operation, so if it is cracked at all, the key needs attention. Check that there are no shiny wear marks other than those on the nose of the key caused by operation of the mechanism, and also that the key still turns easily in the lock and there are no tight spots; all these symptoms indicate extra wear to the lock. If there are multiple keys to a lock, compare them: they will probably differ quite markedly in size and form, having been replaced or added to at various times and with differing degrees of expertise, ranging from the keys which originally came with the lock, to a competent locksmith, to the local hardware shop or blacksmith, to 'Fred the handy church warden' who just happened to have a key which with a bit of bodging would operate the lock. These varied keys will have been causing all sorts of problems to the lock over the years. Some will need to be discarded before they cause further damage but most can be brought to an average and the lock restored to good working order to accommodate them all. If the lock also has a latch mechanism, turn the handle or knob and note how far it has to turn before starting to retract the latch bolt; if it is more than a few degrees, then the latch mechanism must be worn.

***Figure 7*** *The effects of vandalism: a 19th century plate lock from St Michael's Church, Cotham, Nottinghamshire which is in the care of the Churches Conservation Trust before being repaired*

Conservation and care of locks is really a simple matter once an inspection and maintenance programme is instituted and, mostly, is a matter of common sense. For example, a covered escutcheon on the outer face of a south or west facing door can do a lot to inhibit rusting of the mechanism of a lock by keeping out some of the prevailing weather. With woodstock locks check for signs of active woodworm, more common in later, cheaper quality beech stocks, especially if unvarnished, and treat with a proprietary solution. Discourage those who wish to oil locks or pack them with grease; oil and grease goes tacky with time, hampering the smooth working of the mechanism, and grit and dust which enters through the keyhole will stick to it, making a most efficient and wearing grinding paste. If oiling is felt to be absolutely necessary, use a graphite based product. Be aware and do not leave keys, especially attractive ones, easily available, there are light fingered people around. Above all, keep an eye on 'Fred'!

## DOs

- Seek help as soon as a lock becomes awkward or temperamental
- Protect a lock from the weather if possible
- Try new keys with the door open
- Call in a qualified locksmith (consult your Diocesan Advisor or the Master Locksmiths Association)

## DON'Ts

- Use extra force to turn a tight key: the need for more than hand pressure indicates something is wrong
- Use oil or grease in a lock
- Drop keys or toss them down; do treat them gently
- Let well meaning but inexperienced people repair locks or make keys

## FURTHER READING

Books on locks are few and far between, and those available are mostly written with the collector in mind. The most readily obtainable is the small Shire Publications book, *Keys, Their History and Collection* by Eric Monk, priced £5.99, which includes a bibliography. You might be able to borrow *Locks and Keys Throughout the Ages* by Vincent JM Eras, and *An Encyclopaedia of Locks and Builders Hardware*, produced by Josiah Parkes and Sons Limited, from your local library. A short but highly readable history is also provided on the Internet by Chubb at http://www.chubblocks.co.uk/historyoflocks.html. Otherwise it is usually a case of trawling through the second hand and antiquarian bookshops.

The Master Locksmiths Association can be found at: 5D, Great Central Way, Woodford Halse, Daventry, Northants NN11 3PZ
Tel 01327 262255 Fax 01327 262539
E-mail mlahq@locksmiths.co.uk
Website www.locksmiths.co.uk

**VALERIE OLIFENT** is a partner with her husband in their specialist locksmiths business 'The Keyhole'. They specialise in the repair and restoration of pre-1940 locks and the making of appropriate keys. They hold large stocks of period locks, keys and key blanks. The Keyhole, Pilgrim's Progress, Far Back Lane, Farnsfield, Nr Newark, Notts NG2Z 8JX
Tel 01623 882590

The restored organ at Blofield church

# ORGAN DONORS

## Rodney Briscoe

FINE VICTORIAN ORGANS are being transferred from derelict city churches to country parishes, and there are signs that this is a growing trend.

Cities have many large and impressive churches with equally large and impressive organs, but often their congregations are dwindling as people move out, and now many inner city churches are no longer needed for worship. In the last few years several of these organs have been re-located to country churches which could not have afforded such fine instruments in the past, where they replace either small pipe organs or, more often today, defunct electronics.

Examples of such alterations in East Anglia include Great Barton in Suffolk where a good home was found for a Maley, Young and Oldknow organ from Hounslow, removed from the church as the demolition men moved in. At Maldon United Reformed Church a three manual organ from Harrow was installed much to the delight of the enthusiastic congregation. And at Blofield, just east of Norwich, a fine rescued organ from north London was installed.

*The soundboard being restored at the workshops of W&A Boggis*

### ST ANDREW'S & ST PETER'S, BLOFIELD

The story of the Blofield organ is particularly interesting. The church of St Andrew and St Peter is a typically large Norfolk parish church, on the edge of the Broads which had housed, since 1886, a small two manual Norman & Beard organ of simple specification. By 1998 this instrument, although a quality organ in good condition, was deemed to be too small for the church's needs.

It was Rodney Briscoe of W & A Boggis who was contacted by church organist Geoff Sankey when a possible replacement instrument was found in London. A visit to St Peter's, Upper Holloway revealed a once admirable Victorian church that was now sadly deteriorating, and an 1880 William Hill two manual mechanical action organ in poor condition. However it was clear that underneath the dust and neglect there was a well designed and built organ, which needed a good home before the conversion of the building to luxury apartments.

Rodney Briscoe had no hesitation in recommending this organ for Blofield church, knowing from past experience that with skilled restoration it could be successfully rebuilt in a new location. The necessary quotation was produced which was accepted by the Parochial Church Council.

Before an alteration such as the removal of the organ can be made to any church in the ownership of the Church of England, permission must be obtained from the Diocesan Advisory Commission of the Church of England. (Similar requirements exist for all the other main church organisations in the United Kingdom, as described on page 37.) To gain the permission required, which is known as a 'Faculty', applications to each diocese were made with the help of diocesan organ advisors in Norwich and London. Once the formalities were duly completed, arrangements were made for dismantling and collecting the organ from Holloway.

*Detail of the front pipes (left) and the console (right) after restoration*

At the time, Geoff Sankey worked for the famous Adnams brewery in Suffolk, and as their contribution, he arranged for one of their lorries to call at St Peter's Church after delivering in London. So having unloaded some Broadside ale, the lorry took a full broadside of 16ft pedal open diapasons back to Blofield.

Organs in bits take up much more space than when they are assembled, so when it was all back in Diss, the workshop was full of rather dirty organ parts. A thorough examination prior to restoration revealed much of interest, and by referring to the records from the British Institute of Organ Studies (BIOS), the history of the instrument became clearer.

The original order from the factory records of the maker, William Hill, was found through BIOS, and it made interesting reading. It seems that the specification changed several times during the building, which took from August to December 1880. It also shows that Hill made extensive use of reconditioned pipes and other parts. The late Victorian period was a boom time for organ builders in this country, and organs were modified and rebuilt often, sometimes not long after they were built, and the discarded parts were re-used. When dismantling organs of this age and size it is very common to find screwholes and paint marks in unexpected places, showing that the parts had been used before. Hill's order book also showed that neither the mixture nor the trumpet ranks of pipes were fitted at the time. Details of the company's 1931 overhaul indicate that this was when the trumpet was fitted.

At some time in the early 20th century the swell box had been enlarged and a clarionet reed stop had been added. The 1931 overhaul also added to the pedal department, with the addition of a bourdon and bass flute. Furthermore, at this time the pedal action was changed from mechanical to pneumatic. With the current vogue for returning organs to their original specifications, these modifications were carefully considered. It was decided, however, that they were not out of keeping with the character of the organ, and added much to the instrument.

The mixture, however, was a different question. It was decided to fit the mixture that the organ had been waiting for for over 100 years, but this needed a separate faculty. In order to get the faculty, research had to be done to decide the correct composition of the two-rank mixture. Using other Hill organs of the period, a composition of 19–22, breaking back to 15–19 then to 12–15 in the upper octaves was agreed. Fortunately, Rodney Briscoe maintains an extensive stock of good quality pipework, and he was able to make up the mixture from reconditioned pipes of the correct period and voicing.

After the rebuild the specification of the organ is:

**Great:** open diapason 8, hohl flute 8, dulciana 8, principal 4, wald flute 4, fifteenth 2, mixture ii, trumpet 8

**Swell:** gamba 8, clarabella 8, salicional 8, principal 4, flautina 2, cornopean 8, oboe 8, clarionet 8

**Pedal:** open diapason 16, bourdon 16, bass flute 8

**Couplers:** swell to great, swell to pedal, great to pedal

Restoration work began once all the main parts were back in the workshop, including the organ's bellows. These were nearly as big as two double beds and it needed all W&A Boggis's staff of four to lift and move them. The bellows store and pressurise the air that the organ needs. In the modern age efficient electric blowers mean that bellows are much smaller than they used to be, as they no longer have to be big enough to compensate for the fatigue of the man – or boy – pumping the blowing handle. But restoring this instrument meant that the original enormous bellows had to be retained.

The old leather was stripped off and new leather was glued on in its place. Organ builders still use animal-based glue, just as they did when the organ was built. This has several advantages: it can be removed with hot water and it does not react with the leather and wood in an adverse way, as do some chemically manufactured glues. But, as with several aspects of organ building, these old methods and materials are used not out of a desire to be authentic, but simply because there is no better way. The re-leathering work included the feeders as well as the bellows.

## TERMINOLOGY

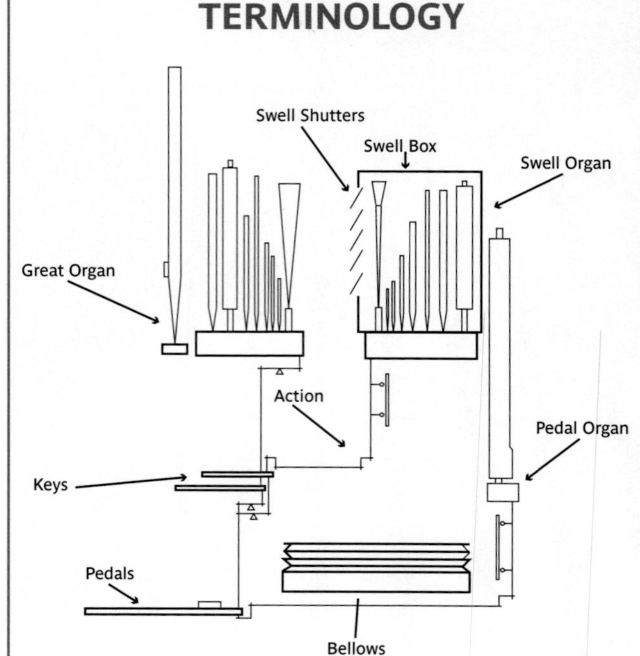


An organ consists of a console with several keyboards and a row of pedals, each connected to a separate section of the instrument (also called 'organ'). Each section contains several sets of pipes – called ranks. Each rank has a different pitch and tone. Starting from the top keyboard, the layout is:

- **the swell organ** – several ranks of pipes, enclosed in a louver-fronted box. These shutters can be moved by the organist through a pedal to increase or decrease the volume.
- **the great organ** – unenclosed ranks of pipes, of louder volume and more strident tone than the swell. The foundation of any organ is the pipe of the great organ, with the swell's more subtle tones providing colour and contrast.
- **the choir organ** – usually enclosed, though not always. Its softer stops are used for accompanying small groups of singers.
- **the pedal organ** – the largest pipes in the organ, often comprised of just one or two ranks of bass pipes.

The ranks are either flue pipes – like a whistle, or reeds – like an oboe or clarinet. The majority of pipes are metal flue pipes, with wood used for softer tones. Stopped pipes produce another variation. A mixture rank consists of several flue pipes tuned to different notes. Reed pipes produce a distinctive tone, with most organs typically having one or two reed ranks.

The ranks are controlled by the stops on the console. Other stops can be used to connect the keyboards to each other, opening wider tone variations.

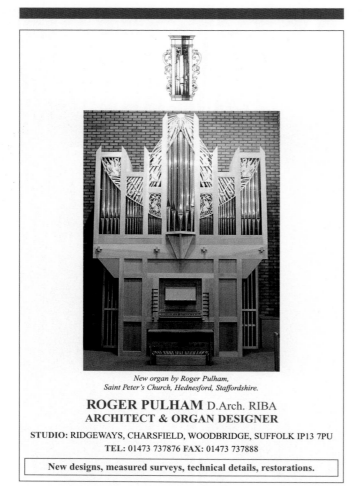

The mechanical action was completely dismantled and cleaned and the iron parts were replaced with new phosphor bronze, which resists the often damp atmosphere of a church better. The soundboards were stripped down and 'flooded', this means sizing it thoroughly with slightly watered down glue to seal any wind leaks between the bars of the soundboard. The pallets were recovered with new felt and leather, the drawstop slides cleaned and re-graphited so that they move easily and cleanly.

The words 'pneumatic action' often seem to strike fear into the hearts of organ players, advisers, and even builders, but a properly restored and maintained pneumatic action is no problem at all. Organ builders without a long heritage in the industry may have never seen pneumatics in good condition and find the vagaries of air pressure harder to understand than a mechanical action. However, Mr William Boggis, who founded the business, had been apprenticed to the masters of pneumatic actions in the heyday of the technology and had himself trained Mr Briscoe. So Mr Briscoe was totally confident in retaining – after careful restoration – the pedal action in this organ.

The oak casework was carefully cleaned and refinished, returning it to its original colour and the delicate stencil work on the pipes, so characteristic of organs of this period, was cleaned. The work is of such good quality that gentle cleaning with soap and water was all that was needed to return the pipes to their former glory, with no retouching necessary.

The bourdon pipes had to be re-located on the other side of the organ for the new position in Blofield church. The casework had to be modified, since the organ was to be facing a different direction – at Holloway it was on the south wall facing north and at Blofield it is the other way around. The pipes had to be changed from one side to the other and a new case panel was made to match the existing panels on the rest of the casework.

Overall this was a most successful project. The instrument fits perfectly with the rest of the church. If Blofield had had more money in the 1880s it is easy to see that they would have had an organ somewhat like this. The church was impressed with Rodney Briscoe's confidence that he could position the organ so that the tallest centre pipe was in the line with the centre of the arch.

This illustrates that there is much that can be done to provide a church with a high quality instrument that comes without the expense, in time as well as money, of a new organ and has the historical integrity and proven reliability of a older instrument. And two churches gained a new organ from this job, as Blofield's former organ was also moved to Colkirk, a village in central Norfolk.

**RODNEY BRISCOE** of W & A Boggis has been organ building for over 40 years, having started at the age of 15. He has gone from apprentice to proprietor of the Diss, Norfolk firm, and recently moved into new purpose-designed premises. He finds satisfaction in the varied aspects of his chosen vocation from tuning small country parish church organs to building new instruments.

# ORGAN BUILDERS

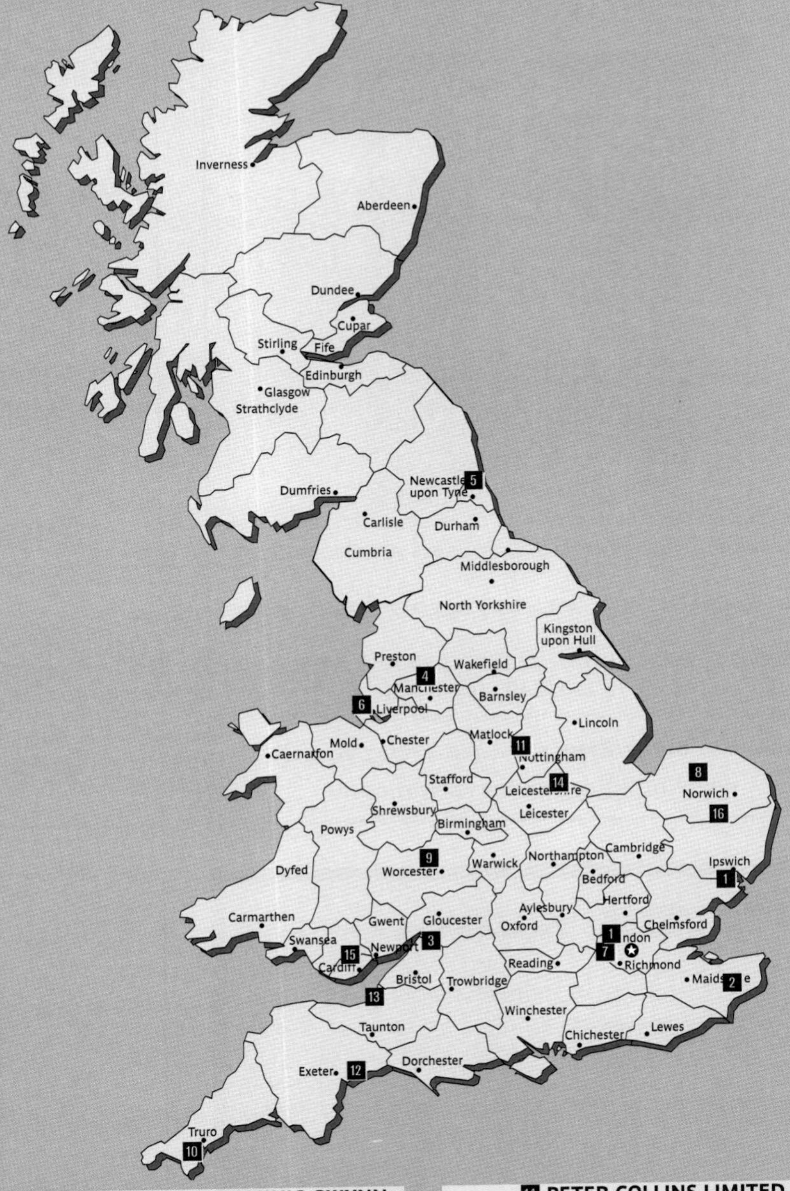


**1 BISHOP & SON**
58, Beethoven Street, London W10 4LG
Tel 020 8969 4328
also at Ipswich 01473 255165
Established 1795. New organs, rebuilds, restoration and tunings. Specialists in restoration of historic instruments.

**2 F H BROWNE & SONS**
The Old Cartwright School, The Street, Ash, Canterbury CT3 2AA
Tel 01304 813146 Fax 01304 812142
Established 1871. Comprehensive range of services available from tunings to new organs. IBO Business Member.

**3 E A CAWSTON**
Hornshill Cottage, Bristol Road, Cam, Dursley, Glos GL11 5JA
Tel/Fax 01453 890720
Small family business specialising in mechanical action, voicing and tonal design. Turning, restoration and overhauls carried out with meticulous care.

**4 GEORGE SIXSMITH & SON LTD**
Hillside Organ Works, Carrhill Road, Mossley, Ashton-under-Lyne, Lancs OL5 0SE
Tel 01457 833009 Fax 01457 835439
Builders of new pipe organs. Restoration and maintenance. Fully comprehensive workshop.

**5 HARRISON & HARRISON LTD**
St John's Road, Meadowfield, Durham DH7 8YH
Tel 0191 378 2222 Fax 0191 378 3388
See advertisement on page 56.

**6 HENRY WILLIS & SONS LTD**
6A, Triumph Park, Speke Hall Road, Liverpool L24 9GQ
Tel 0151 486 1845 Fax 0151 486 1926
Long established company with extensive factory facilities. Branches at: Glasgow, Leeds, Sheffield, Liverpool, Hereford, London. IBO accredited, including historic restoration.

**7 HERITAGE PIPE ORGANS LIMITED**
Arch 133, MacFarlane Road, London W12 7LA
Tel 07958 328818 Fax 020 8614 6802
Rebuilding, restoration and tuning of fine tracker and pneumatic instruments.

**8 HOLMES & SWIFT ORGAN BUILDERS**
Unit 6, The Drift, Fakenham, Norfolk NR21 8EF
Tel/Fax 01328 863400 Mobile 07767 754743
Specialists in restoration, with emphasis on traditional methods and materials. Free initial survey and illustrated report. Regional tuning and maintenance. IBO member.

**9 J BLENEY**
Hill Court, Grafton Flyford, Worcester WR7 4PL
Tel/Fax 01905 391217
A personal service on tunings, maintenance, overhauls, and small rebuilds of pipe organs. Also restorations using traditional methods and materials.

**10 LANCE FOY**
Quartane, Porth Kea, Truro, Cornwall TR3 6AL
Tel/Fax 01872 277585 Mobile 07831 118864
A friendly, dedicated firm offering a complete, high quality organ building service throughout the South and West. IBO accredited.

**11 MARTIN GOETZE & DOMINIC GWYNN**
5 Tan Gallop, Welbeck, Worksop, Nottinghamshire S80 3LW
Tel 01909 485635 Fax 01909 485635
E-mail dominic@goetzegwynn.co.uk
Website www.goetzegwynn.co.uk
Craft workshop specialising in the manufacture of classical English organs using traditional materials and techniques, and the meticulous restoration of historic classical organs.

**12 MICHAEL FARLEY**
2 Kersbrook Farm, Budleigh Salterton, Devon EX9 7AF
Tel 01395 442842 Fax 01395 446680
Mobile 07836 243036
E-mail mjfarley@nicolaj.easynet.co.uk
Restoration, rebuilding, tuning throughout the south of England and beyond. Many fine examples of workmanship available for inspection. IBO Business Member.

**13 PERCY DANIEL & CO LTD**
Beach Avenue, Clevedon, Somerset BS21 7XX
Tel 01275 873273 Fax 01275 342747
New organs, rebuilding, restoration, overhaul and tuning. Located adjacent to M4/M5, giving easy access to all parts of the country.

**14 PETER COLLINS LIMITED ORGAN BUILDERS**
42 Pate Road, Melton Mowbray, Leicestershire LE13 0DG
Tel 01664 410555 Fax 01664 410535
E-mail pd_collins@lineone.net
A company of multi-skilled personnel able to address the exacting demands of tuning, restoring or creating new pipe organs.

**15 PETER HINDMARSH**
Penuel Chapel, Penuel Road, Pentyrch, Cardiff CF15 9QJ
Tel/Fax 029 2089 1151
Specialising in new organs with mechanical action and sensitive restoration of historic instruments.

**16 W & A BOGGIS (R E BRISCOE)**
Roydonian Works, Roydon, Diss, Norfolk IP22 4EQ
Tel/Fax 01379 643599
E-mail rodney.briscoe@talk21.com
Rodney Briscoe offers practical advice and a high standard of workmanship for all types of organs and actions to suit specialist locations.
See editorial article on page 44.

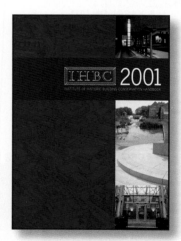

**Advisory Board for Redundant Churches, The**
Cowley House, 9 Little College Street, London SW1P 3XS
Tel 020 7898 1870

**Allchurches Trust Ltd**
Beaufort House, Brunswick Road, Gloucester GL1 1JZ
Tel 01452 528533 Fax 01452 308860

**Ancient Monuments Board for Wales**
National Assembly for Wales, Crown Building,
Cathays Park, Cardiff CF10 3NQ
Tel 029 2082 6376

**Ancient Monuments Society**
St Ann's Vestry Hall, 2 Church Entry, London EC4V 5HB
Tel 020 7236 3934 Fax 020 7329 3677
www.ancientmonumentssociety.org.uk

**Architectural Heritage Fund, The**
Clareville House, 26/27 Oxendon Street,
London SW1Y 4EL
Tel 020 7925 0199 Fax 020 7925 0199
www.ahfund.org

**Architectural Heritage Society of Scotland**
The Glasite Meeting House, 33 Barony Street,
Edinburgh EH3 6NX
Tel 0131 557 0019 Fax 0131 557 0049
www.ahss.org.uk

**Art Loss Register, The**
12 Grosvenor Place, London SW1X 7HH
Tel 020 7235 3393 Fax 020 7235 1652
www.artloss.com

**Association for Studies in the Conservation of Historic Buildings**
Institute of Archaeology, 31-34 Gordon Square,
London WC1H 0PY
Tel 020 7973 3326 Fax 020 7973 3090

**Association of Burial Authorities**
155 Upper Street, Islington, London N1 1RA
Tel 020 7288 2522 Fax 020 7288 2533

**Association of Independent Organ Advisers**
Lime Tree Cottage, 39 Church Street, Haslingfield,
Cambridge CB3 7JE
Tel/Fax 01223 872190 www.aioa.org.uk

**Baptist Building Fund, The**
11 Avening Close, Nailsea, Bristol BS48 4TB
Tel/Fax 01275 795344

**British Artist Blacksmiths Association**
Yew Tree Cottage, Bredenbury, Bromyard,
Herefordshire HR7 4TJ
Tel 01885 482572 www.baba.org.uk

**British Geological Survey**
Keyworth, Nottingham NG12 5GG
Tel 0115 936 3171 Fax 0115 936 3593

**British Metal Casting Association, The**
Boardesley Hall, The Holloway, Alvechurch,
Birmingham B48 7QB
Tel 01527 585222 Fax 01527 590990
www.bmca.org

**British Society of Master Glass Painters**
5 Tivoli Place, Ilkley, West Yorkshire LS29 8SU
Tel/Fax 01943 602521 www.bsmgp.org.uk

**Cadw: Welsh Historic Monuments**
Crown Building, Cathays Park, Cardiff CF10 3NQ
Tel 029 2050 0200 Fax 029 2050 0300
www.cadw.gov.uk

**Capel - Chapels Heritage Society**
c/o RCAHMW, Crown Building, Plas Crug, Aberystwyth,
Ceredigion SY23 1NJ
Tel 01970 621210 Fax 01970 627701
www.rcahmw.org.uk/capel/

**Cathedral Architects Association**
46 St Mary's Street, Ely, Cambridgeshire CB7 4EY
Tel 01353 660660 Fax 01353 660661

**Cathedrals Fabric Commission for England, The**
Church House, Great Smith Street, London SW1P 3BL
Tel 020 7898 1863 Fax 020 7898 1881

**Catholic Bishops Conference of England & Wales**
Liturgy Office, 39 Eccleston Square, London SW1V 1PL
Tel 020 7821 0553 Fax 020 7630 5166
www.liturgy.demon.co.uk

**Chapels Society, The**
1 Newcastle Avenue, Beeston,
Nottinghamshire NG9 1BT
Tel 0115 922 4930
www.britarch.ac.uk/chapelsoc/index.html

**Church Monuments Society, The**
34 Bridge Street, Shepshed, Leicestershire LE12 9AD
Tel 01509 650637

**Church of England, General Synod**
Church House, Great Smith Street, London SW1P 3NZ
Tel 020 7898 1000 Fax 020 7898 1369
www.c-of-e.anglican.org

**Church of Scotland Committee on Artistic Matters**
121 George Street, Edinburgh EH2 4YN
Tel 0131 225 5722 Fax 0131 220 3113

**Church of Scotland, The**
121 George Street, Edinburgh EH2 4YN
Tel 0131 225 5722 Fax 0131 240 2246
www.churchofscotland.org.uk

**Churches Conservation Trust, The**
89 Fleet Street,
London EC4Y 1DH
Tel 020 7936 2285
Fax 020 7936 2284

**Conservation Register, The**
UKIC, 109 The Chandlery,
50 Westminster Bridge Road,
London SE1 7QD
Tel 020 7721 8246

**Corpus Vitrearum Medii Aevi Archive**
National Monuments Record Centre,
Kemble Drive, Swindon,
Wiltshire SN2 2GZ
Tel 01793 414600
Fax 01793 414606
www.english-heritage.org.uk

**Council for British Archaeology**
Bowes Morrell House,
111 Walmgate,
York YO1 2UA
Tel 01904 671417
Fax 01904 671384
www.britarch.ac.uk

**Council for Scottish Archaeology, The**
c/o National Museums of Scotland, Chambers Street,
Edinburgh EH1 1JF
Tel 0131 247 4119
Fax 0131 247 4126
www.britarch.ac.uk/csa/

**Council for the Care of Churches**
Church House,
Great Smith Street,
London SW1P 3NZ
Tel 020 7898 1866
Fax 020 7898 1881

**Department for Transport Local Government and the Regions, The**
Eland House, Bressenden Place, London SW1E 3EB
Tel 020 7944 3000 www.dtlr.gov.uk

**Dúchas - The Heritage Service**
Department of Arts, Heritage Gaeltacht and Islands,
7 Ely Place, Dublin 2, Ireland
Tel +353-1-647 2300 Fax +353-1-662 0283
www.heritageireland.ie

**Ecclesiastical Architects and Surveyors Association**
Property Department, Diocese of Salisbury,
Church House, Crane Street, Salisbury SP1 2QB
Tel 01722 411933 Fax 01722 329833

**Ecclesiological Society, The**
38 Roseberry Drive, New Malden, Surrey KT3 4JS
Tel 020 7213 8724 www.eccl-soc.demon.co.uk

**English Heritage**
23 Savile Row, London W1X 1AB
Tel 020 7973 3000 Fax 020 7973 3001
www.english-heritage.org.uk

**East Midlands Region**
44 Derngate, Northampton NM1 1UH
Tel 01604 735400 Fax 01604 735401

**East of England Region**
62-74 Burleigh Street, Cambridge CB1 1DJ
Tel 01223 582700 Fax 01223 582701

**London Region**
23 Savile Row, London W1S 2ET
Tel 020 7973 3000 Fax 020 7973 3534

**National Monuments Record**
National Monuments Record Centre,
Kemble Drive, Swindon, Wiltshire SN2 2GZ
Tel 01793 414600 Fax 01793 414804

**North East Region**
Bessie Surtees House, 41 Sandhill,
Newcastle upon Tyne NE1 8JF
Tel 0191 261 1585 Fax 0191 261 1130

**North West Region**
Canada House, 3 Chepstow Street, Manchester M1 5FW
Tel 0161 242 1400 Fax 0161 242 1401

**South East Region**
Eastgate Court, 195-205 High Street,
Guildford GU1 3EH
Tel 01483 252000 Fax 01483 252001

**South West Region**
29 Queen Square, Bristol BS1 4ND
Tel 0117 975 0700 Fax 0117 975 0701

**West Midlands Region**
112 Colmore Row, Birmingham B3 3AG
Tel 0121 625 6820 Fax 0121 625 6821

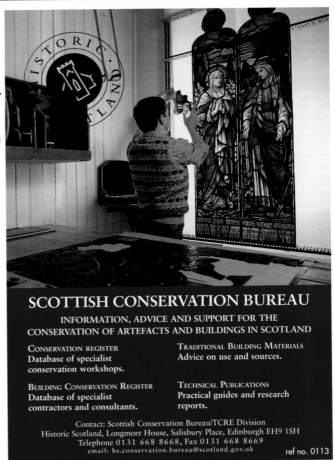



**Yorkshire Region**
37 Tanner Row, York YO1 6WP
Tel 01904 601901 Fax 01904 601999

**Please note:** *English Heritage now includes the former Royal Commission on the Historical Monuments of England (RCHME).*

**English Historic Towns Forum**
PO Box 22, Bristol BS16 1RZ
Tel 0117 975 0459 Fax 0117 975 0460

**ENTRUST**
Acre House, 2 Town Square, Sale,
Cheshire M33 7WZ
Tel 0161 972 0044 Fax 0161 972 0055
www.entrust.org.uk

**Environment & Heritage Service, The**
5/33 Hill Street, Belfast BT1 2LA
Tel 028 9054 3034 Fax 028 9054 3111
www.ehsni.gov.uk

**Friends of Friendless Churches**
St Ann's Vestry Hall, 2 Church Entry,
London EC4V 5HB
Tel 020 7236 3934 Fax 020 7329 3677

**Friends of the City Churches, The**
St Margaret Pattens, Rood Lane,
Eastcheap, London EC3M 1HS
Tel 020 7626 1555 Fax 020 7283 2304
www.london-city-churches.org.uk

**Georgian Group, The**
6 Fitzroy Square, London W1T 5DX
Tel 020 7387 1720 Fax 020 7387 1721
www.heritage.co.uk/georgian/

**Glaziers Trust, The**
Glaziers Hall, 9 Montague Close, London Bridge,
London SE1 9DD
Tel 020 7403 3300 Fax 020 7407 6036

**Heritage Lottery Fund**
7 Holbein Place, London SW1W 8NR
Tel 020 7591 6000 Fax 020 7591 6255
www.hlf.org.uk

**Historic Buildings Council for Wales**
National Assembly for Wales, Crown Building,
Cathays Park, Cardiff CF10 3NQ
Tel 029 2082 6311 Fax 029 2082 6375

**Historic Chapels Trust**
29 Thurloe Street, Kensington, London SW7 2LQ
Tel 020 7584 6072 Fax 020 7225 0607
www.hct.org.uk

## USEFUL ORGANISATIONS

**Historic Churches Preservation Trust**
Fulham Palace, London SW6 6EA
Tel 020 7736 3054 Fax 020 7736 3880

**Historic Scotland**
Longmore House, Salisbury Place, Edinburgh EH9 1SH
Tel 0131 668 8600 Fax 0131 668 8788
www.historic-scotland.gov.uk

**ICOMOS UK (International Council on Monuments & Sites UK)**
10 Barley Mow Passage, Chiswick, London W4 4PH
Tel 020 8994 6477 Fax 020 8747 8464
www.icomos.org/uk/

**Incorporated Society of Organ Builders**
Freepost, Liverpool L3 3AB
Tel/Fax 0151 207 5252
www.musiclink.co.uk/isob/

**Institute of British Organ Building**
63 Colebrooke Row, London N1 8AB
Tel 020 7689 4650 Fax 020 7689 4650
www.ibo.co.uk

**Institute of Field Archaeologists**
The University of Reading, 2 Earley Gate,
PO Box 239, Reading RG6 6AU
Tel 0118 931 6446 Fax 0118 931 6448
www.archaeologists.net

**Institute of Historic Building Conservation**
3 Stafford Road, Tunbridge Wells, Kent TN2 4QZ
***new address and telephone number from 1 January 2002***
Jubilee House, High Street, Tisbury,
Wiltshire SP3 6HA
Tel 01747 873133 www.ihbc.org.uk

**Institution of Civil Engineers, Panel for Historical Engineering Works**
1 Great George Street, London SW1P 3AA
Tel 020 7665 2250 Fax 020 7976 7610
www.ice.org.uk

**International Institute for Conservation of Historic and Artistic Works, The**
6 Buckingham Street, London WC2N 6BA
Tel 020 7839 5975 Fax 020 7976 1564
www.iiconservation.org

**Irish Ecumenical Church Loan Fund**
Inter-Church Centre, 48 Elmwood Avenue,
Belfast BT9 6AZ
Tel 028 9066 3145 Fax 028 9038 1737

**Landfill Tax Credit Scheme** ***– see Entrust***

**London Stained Glass Repository, The**
Glaziers Hall, 9 Montague Close, London Bridge,
London SE1 9DD
Tel 020 7403 3300 Fax 020 7407 6036

**Maintain Our Heritage**
Weymouth House, Beechen Cliff Road,
Bath BA2 4QS
Tel 01225 482074 Fax 0870 137 3805
www.maintainourheritage.co.uk

**Master Carvers Association**
Unit 20, 21 Wren Street, London WC1X 0HF
Tel/Fax 020 7278 8759 www.mastercarvers.net

**Mausolea & Monuments Trust, The**
6 Fitzroy Square, London W1P 6DX
Tel 020 7387 9160 Fax 020 7387 1721

**Methodist Church, Property Committee**
Central Buildings, Oldham Street, Manchester M1 1JG
Tel 0161 236 5194 Fax 0161 236 0752

**Millennium Commission, The**
26th Floor, Portland House, Stag Place,
London SW1E 5EZ
Tel 020 7880 2001 Fax 020 7880 2000
www.millennium.gov.uk

**Monumental Brass Society**
c/o Society of Antiquaries of London,
Burlington House, Piccadilly, London W1V 0HS
Tel 020 8520 5249 Fax 020 8521 8387
www.home.clara.net/williamlack/

**National Archives of Scotland**
HM General Register House, 2 Princes Street,
Edinburgh EH1 3YY
Tel 0131 535 1330 Fax 0131 535 1360
www.nas.gov.uk

**National Assembly for Wales, The**
Crown Building, Cathays Park, Cardiff CF10 3NQ
Tel 029 2050 0200 Fax 029 2082 6375
www.wales.gov.uk

**National Council of Conservation-Restoration**
NCC-R Professional Standards Board
109 The Chandlery,
50 Westminster Bridge Road, London SE1 7QY
Tel 020 7721 8721 Fax 020 7721 8722
www.ukic.org.uk/pacr/

**National Federation of Master Steeplejacks and Lightning Conductor Engineers**
4d St Mary Place, The Lace Market,
Nottingham NG1 1PH
Tel 0115 955 8818 Fax 0115 941 2238
www.nfmslce.co.uk

**National Heritage Memorial Fund**
7 Holbein Place, London SW1W 8NR
Tel 020 7591 6000 Fax 020 7591 6001
www.hlf.org.uk

**National Monuments Record of Scotland**
John Sinclair House, 16 Bernard Terrace,
Edinburgh EH8 9NX
Tel 0131 662 1456 Fax 0131 662 1477
www.rcahms.gov.uk

**National Preservation Office**
The British Library, 96 Euston Road, London NW1 2DB
Tel 020 7412 7612 Fax 020 7412 7796
www.bl.uk/npo/

**National Trust for Scotland, The**
28 Charlotte Square, Edinburgh EH2 4ET
Tel 0131 243 9300 Fax 0131 243 9301

**National Trust, The**
Estates Department, 33 Sheep Street, Cirencester,
Gloucestershire GL7 1RQ
Tel 01285 651818 Fax 01285 657935
www.nationaltrust.org.uk

**National Trust, The**
Head Office, 36 Queen Anne's Gate, Westminster,
London SW1H 9AS
Tel 020 7222 9251 Fax 020 7222 5097
www.nationaltrust.org.uk

**Open Churches Trust, The**
c/o The Really Useful Group Limited,
22 Tower Street, London WC2H 9PW
Tel 020 7240 0880 Fax 020 7240 1204
www.merseyworld.com/faith/html_file/octhead.htm

**Pilgrim Trust, The**
Cowley House, 9 Little College Street,
London SW1P 3XS
Tel 020 7222 4723 Fax 020 7976 0461

**Professional Accreditation of Conservator-Restorers** ***– see National Council for Conservation-Restoration***

**Public Record Office**
Kew, Richmond, Surrey TW9 4DU
Tel 020 8392 5200 Fax 020 8878 8905
www.pro.gov.uk

**Resource: The Council for Museums, Archives and Libraries**
16 Queen Anne's Gate, Westminster,
London SW1H 9AA
Tel 020 7273 1444 Fax 020 7273 1404
www.resource.gov.uk

**Royal Commission on the Ancient and Historical Monuments of Scotland**
John Sinclair House, 16 Bernard Terrace,
Edinburgh EH8 9NX
Tel 0131 662 1456 Fax 0131 662 1477
www.rcahms.gov.uk

**Royal Commission on the Ancient and Historical Monuments of Wales**
Crown Building, Plas Crug, Aberystwyth,
Ceredigion SY23 1NJ
Tel 01970 621200 Fax 01970 627701
www.rcahmw.org.uk

**Royal Incorporation of Architects in Scotland**
15 Rutland Square, Edinburgh EH1 2BE
Tel 0131 229 7545 Fax 0131 228 2188
www.rias.org.uk

**Royal Institute of British Architects Conservation Group**
66 Portland Place, London W1N 4AD
Tel 020 7580 5533 Fax 020 7255 1541

**Royal Institution of Chartered Surveyors, Building Conservation Group**
12 Great George Street, Parliament Square,
London SW1P 3AD
Tel 020 7222 7000 Fax 020 7222 9430
www.rics.org.uk

**Royal Society of Architects in Wales**
Bute Buildings, King Edward VII Avenue,
Cathays Park, Cardiff CF10 3NB
Tel 029 2087 4753 Fax 029 2087 4926

**Royal Society of Ulster Architects**
2 Mount Charles, Belfast BT7 1NZ
Tel 028 9032 3760 Fax 028 9023 7313

**SAVE Britain's Heritage**
70 Cowcross Street, London EC1M 6EJ
Tel 020 7253 3500 Fax 020 7253 3400
www.savebritainsheritage.org

**SAVE Europe's Heritage**
70 Cowcross Street, London EC1M 6EJ
Tel 020 7253 3500 Fax 020 7253 3400
www.savebritainsheritage.org

**Scottish Catholic Archives**
16 Drummond Place, Edinburgh EH3 6PL
Tel 0131 556 3661

**Scottish Catholic Historical Association**
c/o Department of History,
The University of Edinburgh,
William Robertson Building,
50 George Square, Edinburgh EH8 9JY

**Scottish Churches Architectural Heritage Trust**
15 North Bank Street, Edinburgh EH1 2LP
Tel 0131 225 8644 Fax 0131 220 0597

**Scottish Conservation Bureau**
Historic Scotland, Longmore House,
Salisbury Place, Edinburgh EH9 1SH
Tel 0131 668 8668 Fax 0131 668 8669
www.historic-scotland.gov.uk

**Scottish Executive Development Department**
Development Department Secretariat, Area 3-H,
Victoria Quay, Edinburgh EH6 6QQ
Tel 0131 244 0763 Fax 0131 244 0785
www.scotland.gov.uk

**Scottish Record Office** ***– see National Archives of Scotland***

**Scottish Redundant Churches Trust, The**
14 Long Row, New Lanark ML11 9DD
Tel 01555 666023 Fax 01555 665738

**Scottish Society for Conservation and Restoration**
Chauntston, Tartraven, Bathgate Hills,
West Lothian EH48 4NP
Tel 01506 811777 Fax 01506 811888
www.sscr.demon.co.uk

**Society for Church Archaeology, The**
Council for British Archaeology, Bowes Morrell House,
111 Walmgate, York YO1 2UA

**Society for the Protection of Ancient Buildings**
37 Spital Square, Spitalfields, London E1 6DY
Tel 020 7377 1644 Fax 020 7247 5296
www.spab.org.uk

**Society for the Protection of Ancient Buildings in Scotland, The**
The Glasite Meeting House, 33 Barony Street,
Edinburgh EH3 6NX
Tel/Fax 0131 557 1551 www.spab.org.uk

**Society of Architectural Historians of Great Britain**
115 Henderson Row, Edinburgh EH3 5BB
www.bath.ac.uk/centres/casa/sahgb.html

**Stained Glass Museum**
The Chapter House, Ely Cathedral, Ely,
Cambridgeshire CB7 4DN
Tel 01353 660347 Fax 01353 665025
www.stainedglassmuseum.org.uk

**Strict Baptist Historical Society, The**
38 French's Avenue, Dunstable, Bedfordshire LU6 1BH
Tel 01582 602242
www.strictbaptisthistory.org.uk

**Twentieth Century Society, The**
70 Cowcross Street, London EC1M 6EJ
Tel 020 7250 3857 Fax 020 7251 8985
www.c20society.demon.co.uk

**UKIC (United Kingdom Institute for Conservation of Historic and Artistic Works)**
109 The Chandlery, 50 Westminster Bridge Road,
London SE1 7QY
Tel 020 7721 8721 Fax 020 7721 8722
www.ukic.org.uk

**Ulster Architectural Heritage Society**
66 Donegall Pass, Belfast BT7 1BU
Tel 028 9055 0213 Fax 028 9055 0214
www.uahs.co.uk

**Unitarian and Free Christian Churches (General Assembly of)**
Unitarian HQ, Essex Hall, 1-6 Essex Street,
London WC2R 3HY
Tel 020 7240 2384 Fax 020 7240 3089
www.unitarian.org.uk

**Victorian Society, The**
1 Priory Gardens, Bedford Park, London W4 1TT
Tel 020 8994 1019 Fax 020 8995 4895
www.victorian-society.org.uk

**Welsh Church Fund**
Community Grants & External Funding,
Community Development,
Carmarthenshire County Council, Town Hall, Llanelli,
Carmarthenshire SA15 3AH
Tel 01554 742182

**Wesley Historical Society**
34 Spiceland Road, Northfield, Birmingham B31 1NJ
Tel 0121 475 4914

**Working Party on Jewish Monuments in the UK and Ireland, The**
c/o Jewish Memorial Council, 25 Enford Street,
London W1H 2DD
Tel 020 7636 0776

**World Monuments Fund in Britain**
2 Grosvenor Gardens, London SW1W 0DH
Tel 020 7730 5344 Fax 020 7730 5355
www.worldmonuments.org

**Worshipful Company of Glaziers and Painters of Glass, The**
Glaziers Hall, 9 Montague Close, London Bridge,
London SE1 9DD
Tel 020 7403 3300 Fax 020 7407 6036

## INDEX



## ADVISORY ORGANISATIONS

### SCOTTISH CONSERVATION BUREAU / HISTORIC SCOTLAND

*See advertisement on page 49.*

## ANTIQUE & DECORATIVE LIGHTING

### DERNIER & HAMLYN

*See advertisement on page 39.*

## ARCHAEOLOGISTS

### ARCHITECTURAL ARCHAEOLOGY

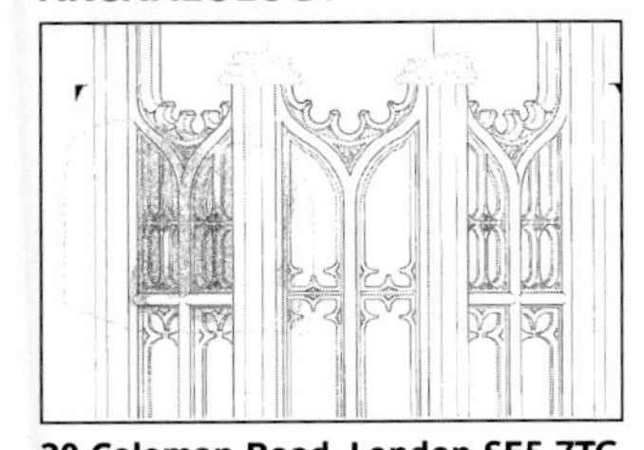

**20 Coleman Road, London SE5 7TG**
**Tel 020 7703 7519**
**E-mail mwsamuel@aol.com *or* klhamlyn@aol.com**
**Mark Samuel PhD, MIFA**

Architectural Archaeology specialises in medieval ecclesiastical architecture, providing archaeological evaluation, rapid analysis, definitive reports and reconstruction drawings of masonry architecture and moulded detail. Reports to MAP2/PPG15 standards. Writing and editing services available. Current clients include the Dean & Chapter of St Paul's Cathedral.

## PRODUCTS AND SERVICES

### IFA YEARBOOK

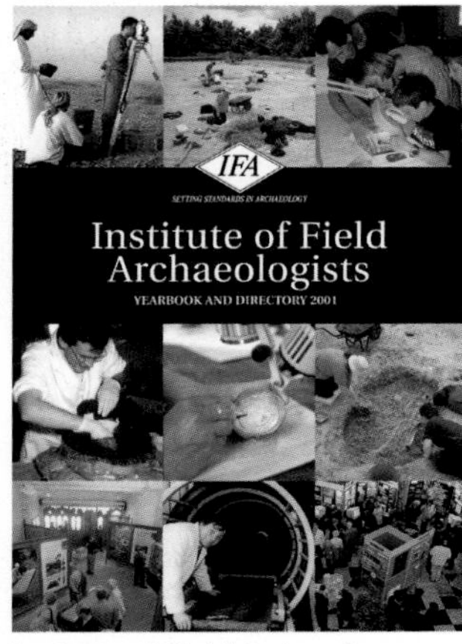


**Institute of Field Archaeology**
**University of Reading,**
**2 Earley Gate, PO Box 239,**
**Reading RG6 6AU**
**Tel 0118 931 6446**
**Fax 0118 931 6448**
**www.archaeologists.net**

Published by Cathedral Communications Limited, this very useful annual working guide to the UK's archaeology profession includes complete listings of all IFA members along with essential industry contacts. Editorial articles highlight the most pressing issues in archaeology, and a directory of specialist services enables you to locate the right archaeologist or support service provider for the job. £15.00, 104 pages.

### OXFORD ARCHAEOLOGICAL UNIT

**Janus House, Osney Mead,**
**Oxford OX2 0ES**
**Tel 01865 243888**
**Fax 01865 793496**
**www.oau-oxford.com**

Historic church specialists. Over 25 years experience in building recording, archaeological evaluation, excavation and historical research. We provide clear and effective advice for all faculty applications. Clients: English Heritage, National Trust, Historic Royal Palaces and churches nationwide.

## ARCHITECTS

### ACANTHUS CLEWS ARCHITECTS LTD

**The Old Swan, Swan Lane,**
**Great Bourton, Banbury,**
**Oxon OX17 1QR**
**Tel 01295 758101**
**Fax 01295 750387**
**E-mail architects@acanthusclews.co.uk**
**www.acanthusclews.co.uk**

The practice is committed to providing quality of design with a high standard of service to clients. We have expertise in repair and conservation of historic buildings, in particular churches and cathedrals on which we act as architects to both Llandaff and Coventry Cathedrals. We seek to combine the advantages of modern technology with those of proven methods to achieve excellence.

### AMS JOHN ARCHITECTS

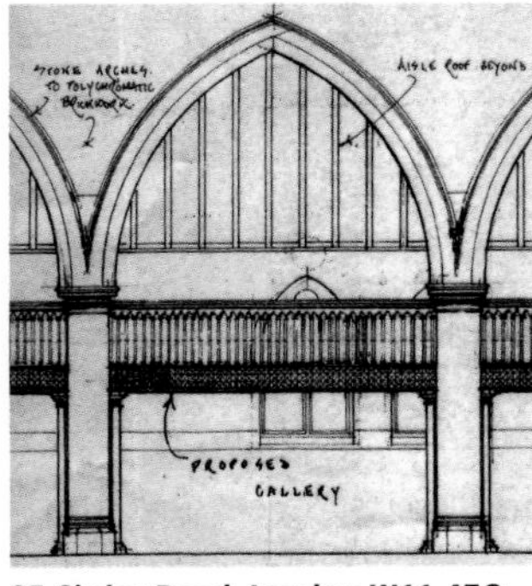


**95 Sirdar Road, London W11 4EQ**
**Tel 020 7727 7551**
**Fax 020 7243 2525**
**E-mail murray@amsjohn.demon.co.uk**

Murray John who is an inspecting architect for the London diocese formed this imaginative young practice in 1980 offering a rare combination of the sensitive repair of historic buildings and fresh, high quality design. AMS John takes particular care to understand the special requirements of church communities, and to plan for the long-term well being of their buildings.

### BYROM CLARK ROBERTS LTD

**117 Portland Street,**
**Manchester M1 6EH**
**Tel 0161 236 9601**
**Fax 0161 236 8675**
**Contact Andrew Hawksworth**
**▶ Jubilee House, West Bar Green,**
**Sheffield S1 2BT**
**Tel 0114 275 7879**
**Fax 0114 272 8954**
**Contact Alex Roberts**
**E-mail bcrmcr@bcr.uk.com**
**www.bcr.uk.com**

A multi-disciplinary practice of architects, building surveyors and structural engineers with 75 years experience of ecclesiastical work providing a balance between innovative thinking and the conservation of historic building character. The firm specialises in conservation, repair, reordering and sensitive extensions and carries out 60 regular quinquennial inspections across eight dioceses and several Methodist regions working regularly with English Heritage in grant-aided and other projects.

### CARDEN & GODFREY ARCHITECTS

**9 Broad Court, Long Acre,**
**London WC2B 5PY**
**Tel 020 7240 0444**
**Fax 020 7836 2244**
**E-mail vcab@cardenandgodfrey.demon.co.uk**

Specialists in all aspects of historic architecture: conservation, repairs, new buildings in sensitive areas, interior design, etc. Clients include the National Trust, English Heritage, Oxford colleges, cathedrals and churches.

## CHRIS ROMAIN ARCHITECTURE

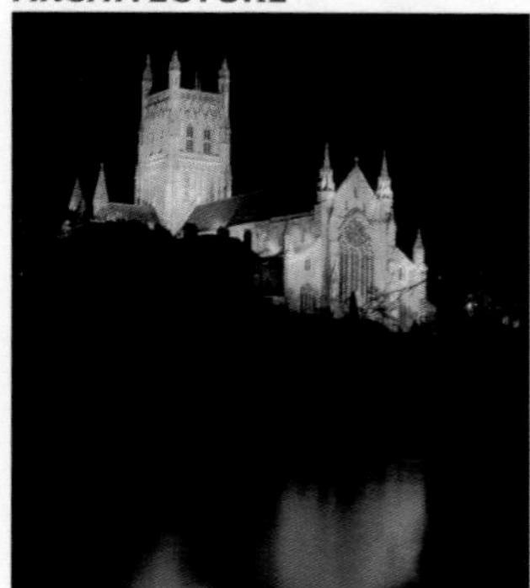

**Griffin Mews, 22 High Street, Fordingbridge, Hampshire SP6 1AX**
**Tel 01425 650980**
**Fax 01425 650978**
**E-mail chris.romain@virgin.net**
**www.chrisromain.co.uk**
Chris Romain Architecture is a small practice specialising in ecclesiastical and historic work across the South of England to the West Midlands. Chris Romain is Cathedral Architect at Worcester and Architect to Bath, Sherborne and Milton Abbeys. Alterations and conservation of historic commercial buildings are also undertaken.

## CLAGUE

**62 Burgate, Canterbury, Kent CT1 2BH**
**Tel 01227 762060**
**Fax 01227 762149**
**▶ 13 North Street, Ashford, Kent TN24 8LF**
**Tel 01233 624354**
**Fax 01233 610018**
Clague specialise in ecclesiastical and conservation architecture. Established in 1936 and now a team of 30 professionals. Appointed to many parish churches and The Dean and Chapter of Canterbury Cathedral (in respect of secular properties). Clague enjoy regularly carrying out quinquennial inspections, repairs, reordering, lighting and heating schemes. Brochure available.

## DAVID PITTS CHARTERED ARCHITECTS

**12a The Waits, St Ives, Cambs PE27 5BY**
**Tel 01480 466213**
**Fax 01480 493330**
**E-mail dpittsriba@aol.com**
**▶ PO Box 8, Oundle, Northants PE8 5JQ**
**Tel 01780 470170**
**Fax 01780 470800**
Practice established 1976, approved architect to Ely, Lincolnshire, Leicestershire and Peterborough DACs for quinquennial inspections. Specialists in the conservation and repair of fabric to ecclesiastical buildings, alterations and reordering.

## DONALD INSALL ASSOCIATES

**19 West Eaton Place, London SW1X 8LT**
**Tel 020 7245 9888**
**Fax 020 7235 4370**
**▶ Canterbury Tel 01227 761616**
**▶ Cambridge Tel 01223 303111**
**▶ Chester Tel 01244 350063**
**▶ Shrewsbury Tel 01743 351244**
**▶ Bath Tel 01225 469898**
The practice has over 40 years experience in the care of historic buildings and towns, including the inspection, repair and re-ordering of churches throughout the country. Its members are currently approved as inspecting architects by the Dioceses of Canterbury, Chelmsford, Chichester, Durham, Lichfield, Lincoln, London, Norwich, Oxford, Peterborough, Southwark and St Albans.

## EDWARD SARGENT CONSERVATION ARCHITECT

**Heritage House, 79-80 High Street, Gravesend, Kent DA11 0BH**
**Tel 01474 535221**
**Fax 01474 564857**
Small architectural practice specialising in repairs and alterations to historic buildings and the design of new buildings in historic environments.

## HAIGH ARCHITECTS

**29 Lowther Street, Kendal, Cumbria LA9 4DH**
**Tel 01539 720560**
**Fax 01539 723570**
**E-mail haigharch@aol.com**
**www.haigharchitects.co.uk**
We provide a sensitive and skilled approach to the repair and conservation of churches, including contextual reordering and extensions, lighting and heating schemes. Quinquennial inspections for Diocese of Carlisle, Churches Conservation Trust and United Reformed Church.

## HAWKES EDWARDS & CAVE

**1 Old Town, Stratford-upon-Avon, Warwickshire CV37 6BG**
**Tel 01789 298877**
Over 45 years of experience in dealing with listed historic buildings and churches, where a sensitive, skilled and creative approach is required, both in repair and alteration. They act, amongst others, for the Churches Conservation Trust and the Ancient Monuments Society.

## MARGARET & RICHARD DAVIES AND ASSOCIATES

*See advertisement on page 15.*

## MATTHEW THOMAS CHARTERED ECCLESIASTICAL ARCHITECT

**25 Cathedral Street, Norwich, Norfolk NR1 1LU**
**Tel 01603 629469**
The practice, approved in five dioceses, undertakes repairs and new design. A comprehensive, efficient and personal service is provided, with reasonable fees. Winner of 2001 Presidents' Award.

## THE NYE SAUNDERS JOHN DEAL PRACTICE

**3 Church Street, Godalming, Surrey GU7 1EQ**
**Tel 01483 418600**
**Fax 01483 418655**
**E-mail architecture@nsjdp.co.uk**
35 years experience in care and design for historic and new churches; quinquennial inspections undertaken in Guildford, Southwark and London Dioceses. Repairs, re-ordering, extensions and design of liturgical furniture.

## PETER CODLING ARCHITECTS

**7 The Old Church, St Matthews Road, Norwich, Norfolk NR1 1SP**
**E-mail pcodling@globalnet.co.uk**
**Tel 01603 660408**
**Fax 01603 630339**
Church repairs, reordering and extensions; quinquennial reports. Repair and conversion of buildings of all ages and types. Housing for individual clients and special needs groups.

## ROBERT SEYMOUR CONSERVATION

**The Merchants House, 10 High Street, Totnes, Devon TQ9 5RY**
**Tel 01803 865568**
**Fax 01803 834722**
Conservation architects and historic building consultants. The practice has over 25 years experience carrying out sympathetic, appropriate repairs to a wide range of historic and listed buildings. It has strong links with English Heritage and SPAB, working with private clients, local authorities, charitable trusts, almshouse associations, churches and other groups. With other offices in Dartmouth and London, the firm's architects, qualified and experienced in building conservation, are able to carry out detailed surveys, evaluations and repair programmes. Robert Seymour Conservation combines technical expertise with innovative and sensitive design by working with a network of specialists as well as providing sensitive solutions to new or historic building projects, often in sensitive urban conservation areas, throughout the south of England.

## ROGER MEARS ARCHITECTS

**2 Compton Terrace, London N1 2UN**
**Tel 020 7359 8222**
**Fax 020 7354 5208**
Founded in 1980, the practice has built a reputation for sensitive work to historic buildings, both domestic and ecclesiastical, guided by the principles of SPAB. Past work includes repairs to St Simon and St Jude, Llanddeusant, Carmarthenshire *(illustrated)*, Emmanuel Church, Forest Gate, London and listed houses in and around London.

## ROGER PULHAM

*See advertisement on page 46.*

## SIMPSON & BROWN

**St Ninian's Manse, Quayside Street, Edinburgh EH6 6EJ**
**Tel 0131 555 4678**
**Fax 0131 553 4576**
**E-mail admin@ simpsonandbrown.co.uk**
**www.simpsonandbrown. co.uk**
Repairs and alterations to churches have been an important part of Simpson & Brown's work since the practice was founded in 1977. They have been consultants to over 70 churches of all denominations throughout Scotland and Northern England. Their work is based on thorough initial inspection which forms the basis of a scheme of repair.

## STAINBURN TAYLOR ARCHITECTS

**Bideford House, Church Lane, Ledbury, Herefordshire HR8 1DW**
**Tel 01531 634848**
**Fax 01531 633273**
**E-mail architects@ stainburntaylor.co.uk**
This award winning practice has been responsible for major conservation and refurbishment work on numerous churches in Herefordshire, Worcestershire and Shropshire and are architects to Gloucester Cathedral. The practice has developed considerable expertise in the diagnosis and conservation of buildings both ecclesiastical and secular.

## THOMAS FORD & PARTNERS
### Chartered Architects & Surveyors

**177 Kirkdale, Sydenham, London SE26 4QH**
**Tel 020 8659 3250**
**Fax 020 8659 3146**
**E-mail tfp@thomasford.co.uk**
Thomas Ford & Partners is an established firm of architects and surveyors with 75 years experience in all aspects of ecclesiastical work including new churches, remodelling, extensions, alterations, maintenance and conservation. The practice holds appointments as Inspecting Architects for over 250 churches of all ages and denominations.

## THE VICTOR FARRAR PARTNERSHIP

**57 St Peter's Street, Bedford MK40 2PR**
**Tel 01234 353012**
**Fax 01234 363473**
Chartered architects, established 1962. Everyone trained in modern conservation techniques. Work includes repair, conservation, sensitive extension, refurbishment and reordering. A regular supply of good craftsmen available to enable competitive tendering, with quality. Quinquennial inspections. Consultations.

## THE WHITWORTH CO-PARTNERSHIP
### with Boniface Associates

**18 Hatter Street, Bury St Edmunds, Suffolk IP33 1NF**
**Tel 01284 760421**
**Fax 01284 704734**
**E-mail Whitcp@globalnet.co.uk**
Chartered architects and building surveyors. Modern design solutions and reorderings. We are consultants to all denominations and some local authorities. We are experienced in the diagnosis of building defects, quinquennial inspection reports, sensitive structural repairs and the conservation of building elements and materials. Associated practice in London serving the Home Counties.

# ARCHITECTURAL METALWORK

## CASTING REPAIRS

*See advertisement on page 27.*

## DOROTHEA RESTORATIONS LTD

**Riverside Business Park, St Anne's Road, Bristol BS4 4ED**
**Tel 0117 971 5337**
**Fax 0117 977 1677**
**▶ New Road, Whaley Bridge, High Peak, Derbyshire SK23 7JG**
**Tel 01663 733544**
**Fax 01663 734521**
Established in 1974, the company specialises in the conservation and reinstatement of historic metalwork using traditional methods and authentic materials, including genuine wrought iron, cast iron, bronze and zinc. Recent contracts include medieval ironwork at Salisbury Cathedral, and gates and railings at St Albans Cathedral, St Mary's Newark, Christ Church Spitalfields, and St Michael's Abbey Farnborough.

## SMITH OF DERBY

**112 Alfreton Road, Derby DE21 4AU**
**Tel 01332 345569**
**Fax 01332 290642**
**E-mail sales@smithofderby.com**
**www.smithofderby.com**
Specialist metalwork for new buildings and refurbishment schemes. Work includes special projects for architects, interior designers and churches. All work is done to commission, and examples include balustrades, hand beaten copper weathervanes, polished brass interior fittings, lighting units and stainless steel furnishing items.

# ARCHITECTURAL TERRACOTTA

## LAMBS TERRACOTTA & FAIENCE

*See advertisement on page 39.*

# BRICK SUPPLIERS

## THE BULMER BRICK & TILE CO LTD

**Bulmer, near Sudbury, Suffolk CO10 7EF**
**Tel 01787 269232**
**Fax 01787 269040**
Specialists in producing bricks, terracotta and washed red rubbers for restoration and conservation. Using the finest London bed clays and hand moulding techniques replicating bricks produced on the site from the 15th century. We also provide a full cutting service for specials cut from our own or customers' materials.

# BUILDING CONTRACTORS

## BERNARD A SHEPHERD LTD

**33A Cumberland Street, Macclesfield, Cheshire SK10 1DD**
**Tel 01625 432477**
**Fax 01625 432488**
The company has over 30 years experience in timber frame restoration projects, difficult masonry schemes, restoration of historic building contracts such as churches, schools and residential developments. Clients are Church Conservation Trust, English Heritage, the National Trust, Lloyds Bank etc. The company operates mainly in the North West of England.

## BURLEIGH STONE CLEANING & RESTORATION CO LTD

**The Old Stables, 56 Balliol Road, Bootle, Merseyside L20 7EJ**
**Tel 0151 922 3366**
**Fax 0151 922 3377**
**E-mail info@ burleighstone.freeserve.co.uk**
**www.burleighstone.co.uk**
*See advertisement on page 19.*

## PRODUCTS AND SERVICES

### C R CRANE & SON LTD

**Manor Farm, Main Road, Nether Broughton, Leics LE14 3HB**
**Tel 01664 823366**
**Fax 01664 823534**
**www.crcrane.co.uk**

Chartered builders and churchwrights established in 1910 by the current managing director's grandfather. Specialists in traditional repairs using SPAB and English Heritage methods. Apprentice trained craftsmen experienced in timber framing, brickwork, masonry, leadwork, lime mortars and plasters, backed up by CIOB qualified staff. In-house workshop specialising in traditional and ecclesiastical joinery.

### CONSERVATRIX
David Alcock & Philip Scorer Associates

**▶ David Alcock, 18 Hatherley Street, Cheltenham, Glos GL50 2TT**
**Tel/Fax 01242 702525**
**▶Philip Scorer, 24 Elton Road, Bristol BS7 8DD**
**Tel 01179 257004**
**www.conservatrix.co.uk**

Conservatrix provides conservation and rope-access expertise for historic buildings. Committed to the highest standards of work and business practice, the principals are professional mortar, masonry, and timber conservators with many years practical experience of traditional materials and methods. Rope-access can cut scaffold costs and bring flexibility to project timing.

### HUGH HARRISON CONSERVATION

*See advertisement on page 28.*

### LONGLEY
A division of Kier Regional Ltd

**East Park, Crawley, West Sussex RH10 8EU**
**Tel 01273 561212**
**Fax 01273 564333**

Building restoration and conservation in the central Home Counties. Longley has over 130 years experience working to maintain the nation's heritage buildings. Appropriate skills and technical expertise have been used on projects such as the Hampton Court Palace restoration following the tragic fire. Contracts range from £100,000 to £5 million and can benefit from Wallis joinery and stonework as Longley is also part of the Wallis division within the Kier Group.
*See also Wallis advertisement on inside front cover.*

### MAYSAND LIMITED

*See advertisement on page 15.*

### S & J WHITEHEAD LTD

*See advertisement on page 16.*

### ST BLAISE LTD

*See advertisement on page 24.*

### T J EVERS LTD

**New Road, Tiptree, Colchester, Essex CO5 0HQ**
**Tel 01621 815787**
**Fax 01621 818085**
**E-mail office@tjevers.co.uk**

Established 1918, a traditional contractor in the conservation and restoration of historic buildings offering high calibre craft skills, traditional and modern construction techniques, coupled with quality management. The specialist joinery division produces period and bespoke joinery to the highest standards. Varied clients include English Heritage. Both large and small contracts undertaken.

### TAYLOR DALTON, HERITAGE BUILDING CONTRACTORS

*See advertisement on page 31.*

### WALLIS

**Broadmead Works, Hart Street, Maidstone, Kent ME16 8RE**
**Tel 01622 690960**
**Fax 01622 693553**

Wallis, part of the Kier Group, has earned an exceptional reputation for the quality and excellence of their restoration and refurbishment work and have carried out work in Christ Church Isle of Dogs, St Paul's Hammersmith, St Paul's Knightsbridge, The London Oratory, St Nicholas Chislehurst, Hampton Court Palace, Danson House, Eltham Palace and Ightham Mote.
*See also advertisement on inside front cover.*

### WILLIAM LANGSHAW & SONS

*See advertisement on page 28.*

## CAST STONE

### HADDONSTONE LIMITED

**The Forge House, East Haddon, Northampton NN6 8DB**
**Tel 01604 770711**
**Fax 01604 770027**
**E-mail info@haddonstone.co.uk**
**www.haddonstone.co.uk**

Haddonstone is the leading manufacturer of standard and custom-made reconstructed stonework. The company can replicate original architectural stone elements in practically any shape or size, providing a sensitive and cost effective solution for churches and other historic buildings. A full colour 156 page catalogue is available. Offices also in California, Colorado and New Jersey.

## CLOCKS

### THE CUMBRIA CLOCK COMPANY

**Dacre, Penrith, Cumbria CA11 0HH**
**Tel/Fax 01768 486933**

Full restoration of mechanical movements, carillon and tune playing machines; manufacture and installation of automatic-winding systems, and full restoration of dials. New clock systems and dials also manufactured, complete with striking and chiming mechanisms. Clients include cathedrals, National Trust, and many churches and local authorities. Free quotations given.

### SMITH OF DERBY

**112 Alfreton Road, Derby DE21 4AU**
**Tel 01332 345569**
**Fax 01332 290642**
**E-mail sales@smithofderby.com**
**www.smithofderby.com**

Smith of Derby has since 1856 built, installed and maintained clocks on churches. Specialist skills include restoration, automatic winding and dial regilding using the finest quality materials. Regional centres provide clock maintenance, and year round cover is available for site visits, faultfinding, replacement of hammer wires, weight lines and time motors.


CATHEDRAL COMMUNICATIONS LIMITED

## THWAITES & REED LIMITED (Est 1740)

**Unit 2, Burgess Road, Hastings TN35 4NR**
**Tel 01424 423537 / 01424 431740**
**Fax 01424 431672**
**E-mail thwaitesandreed@aol.com**
**thwaites-reed.co.uk**

Restoration and conservation, new clocks, electric drives, autowinds, night silencers, compliance. Working for PCCs, contractors, architects and diocesan advisors.

# COURSES & TRAINING

## BUILDING CRAFTS COLLEGE

*See advertisement on page 48.*

# DAMP & TIMBER DECAY

## MAYSAND LIMITED

*See advertisement on page 15.*

## ROBINSONS PRESERVATION LIMITED

*See advertisement on page 28.*

## TERMINIX LIMITED

**Heritage House, 234 High Street, Sutton, Surrey SM1 1NX**
**Tel 020 8661 6600**
**Fax 020 8642 0677**
**Branches – 0800 789500**

Specialists in the treatment of damp, woodworm and dry rot with extensive experience on historic and listed buildings. Unique transfusion system of damp proofing, proven over 30 years and BBA certificated. Expertise in epoxy resin techniques for cost effective beam end repairs. Skilled craftsmen and surveyors. BSI registered. Branches nationwide.

# DECORATIVE & STAINED GLASS

## CLIFFORD G DURANT & SON STAINED GLASS CONSERVATION

**The Glasshouse Studio, New Street, Horsham, West Sussex RH13 5DU**
**Tel 01403 264607**
**Fax 01403 254128**

Established 1972, specialising in conservation and restoration of stained glass and leaded lights. Accredited to categories 1, 2, 3 and 4 by the United Kingdom Institute for Conservation of Historic and Artistic Works.

## NICK BAYLISS ARCHITECTURAL GLASS

**152 Warstone Lane, Hockley, Birmingham B18 6NZ**
**Tel/Fax 0121 233 1985**

We design, make and install stained glass windows and leaded lights to the highest standards of craftsmanship. We provide a comprehensive and efficient service, backed by years of experience. Our showroom/workshop is five minutes from the city centre. We also specialise in the restoration of period timber and iron framework.

## NORGROVE STUDIOS

*See advertisement on page 15.*

## OPUS STAINED GLASS

**The Old Village Hall, Mill Lane, Poynings, Brighton, West Sussex BN45 7AE**
**Tel 01273 857223**
**Fax 01273 857161**

Our studio is experienced in the conservation and restoration of churches and historic buildings, and we are specialists in the renovation of windows by 19th century artists. As well as liaising directly with local churches, Opus works closely with restoration architects and conservation organisations including English Heritage.

## RIVERSIDE STUDIO

**9b Curzon Street, Kingston upon Hull, East Yorkshire HU3 6PH**
**Tel 01482 563742**
**Fax 01482 575900**

Specialist conservation, restoration and repair of stained and leaded glazing. New commissions undertaken for stained glass and leaded glazing. Manufacturers of stainless steel wire guards and casements. Conservation, restoration, repair and manufacture of ironwork. Condition reports and surveys carried out. All work carried out nationally. Accredited members of UKIC.

## THE STAINED GLASS SPECIALIST

**35 Leigh Lane, Wimborne, Dorset BH21 2PW**
**Tel 01202 882208**
**Fax 01202 882208**
**www.lead-windows.co.uk**

Steve Sherriff heads this family run company; reliable and experienced in the restoration and conservation of existing glass, lead and frames in all situations. New commissions are welcomed both in the traditional and modern styles. Protective window guards are also supplied and installed. All workmanship and materials to the highest standards.

# FINE ART

## FLEUR KELLY

*See advertisement on page 10.*

# FINE JOINERY

## C R CRANE & SON LTD

**Manor Farm, Main Road, Nether Broughton, Leics LE14 3HB**
**Tel 01664 823366**
**Fax 01664 823534**
**www.crcrane.co.uk**

*See entry in Building Contractors section, page 54.*

## MOTT GRAVES PROJECTS LTD

**Sampleoak Lane, Chilworth, Guildford GU4 8QW**
**Tel 01483 453326**
**Fax 01483 453325**
**www.mottgraves.co.uk**

This company provides a comprehensive woodworking service for private and professional clients. The directors personally organise and manage each commission, carefully combining proven restoration methods with a sensitive approach to new work. Previous works include the refurbishment of County Hall, Westminster and conservation works within Ingress Abbey, Dartford. MGP also designs and manufactures high quality bespoke furniture.

*See advertisements on pages 28 and 32.*

## TANKERDALE LTD

**Johnson's Barns, Waterworks Road, Sheet, Petersfield, Hampshire GU32 2BY**
**Tel 01730 233792**
**Fax 01730 233922**

Tankerdale specialises in the restoration and conservation of joinery, furniture and historic panelling. Established in 1977 and covering all of the UK, they are official advisers to the National Trust and work with churches, museums, architects, contractors and private clients. Recent contracts include conservation of the box pews and panelling at Staunton Harold Church, Leicestershire and the William Hill organ case at Little Houghton, Northampton.

## WALLIS JOINERY

**Broadmead Works, Hart Street, Maidstone, Kent ME16 8RE**
**Tel 01622 690960**
**Fax 01622 693553**

Wallis Joinery, part of the Kier Group, is one of the UK's leading joinery companies offering unparalleled standards of craftsmanship and expertise. They have twice been awarded the prestigious Carpenters Award and are accredited to ISO 9002 and ISO 9001. They provide all components, from standard softwood joinery to complex restoration projects using hardwood, including purpose designed fixtures and fittings in veneers and laminates.

*See also advertisement on inside front cover.*

## HEATING

### CHRISTOPHER DUNPHY ECCLESIASTICAL LTD

**Mitre House, Spotland Stadium North Stand, Willbutts Lane, Rochdale, Lancs OL11 5BE**
**Tel 0800 614100**
**Fax 01706 354815**
**E-mail dunphyheating@zen.co.uk**
Christopher Dunphy Ecclesiastical Ltd is a specialist company dedicated to the design and installation of heating systems exclusively for churches. The company does not manufacture or promote any particular system but will design a system to exactly match the criteria of each church individually. Over 25 years experience. Free surveys.

### J & J W LONGBOTTOM LTD

**Bridge Foundry, Holmfirth, Huddersfield HD7 1AW**
**Tel 01484 682141**
**Fax 01484 681513**
Ornamental gratings for heating duct systems are produced by this long established traditional foundry. In addition to a wide range of standard patterns, castings to match other existing designs can be produced. Catalogue available on request.
*See entry in Roof Drainage section, page 57.*

## HISTORICAL RESEARCH

### SELWOOD AND DUNCAN

**Architectural Survey and Record**
**14 Lambridge, Bath BA1 6BJ**
**Tel/Fax 01225 421550**
**E-mail sd@surveyandrecord.co.uk**
**www.surveyandrecord.co.uk**
*See also entry in Measured Surveys section on this page.*

## INTERIORS CONSULTANTS & CONSERVATORS

### HIRST CONSERVATION

*See advertisement on page 10.*

### HOWELL & BELLION

**66a High Street, Saffron Walden, Essex CB10 1EE**
**Tel 01799 522402**
**Fax 01799 525696**
Specialist church decorators and restorers. Many years experience and knowledge have resulted in a prestigious client list. New schemes undertaken as well as the cleaning and conservation of existing decoration. Projects large or small, interior or exterior. Specialist advice and information available.

### INTERNATIONAL FINE ART CONSERVATION STUDIOS LTD

*See advertisement on page 10.*

### NEVIN OF EDINBURGH

*See advertisement on page 10.*

## LIGHTING CONSULTANTS

### LIGHTING DESIGN & CONSULTANCY

**Newcombe House, 21 Market Place, Wolsingham, Co Durham DL13 3AB**
**Tel/Fax 01388 527809**
**E-mail ldc@ lightingconsultants.co.uk**
**www.lightingconsultants.co.uk**
**Contact Michael Phillips I Eng MILE MSLL ACIBSE Lighting Diploma**
Independent lighting consultants. Founded in 1989, LDC has established extensive experience in interior and exterior lighting design for ecclesiastical and historic buildings of all denominations. Commissions include over 200 church lighting projects, many Grade I listed buildings, works of art and sculptures. Design of bespoke lighting equipment. LDC is registered with the ILE and CIBSE as an independent lighting practice.

## LIGHTING CONTROLS

### EXODUS ELECTRONIC LTD

**Singleton Court, Wonastow Road, Monmouth NP25 5JA**
**Tel 01600 719444**
**Fax 01600 716744**
**E-mail info@exodus-electronic.com**
Lights can now be controlled without unsightly cables running between light fittings and switches. We make a range of wire free lighting controls especially suited to churches and old buildings. These provide on/off and dimming control of incandescent and fluorescent lamps. There is now an alternative to traditional control cabling that is cost-effective, simple to install, flexible, reliable and attractive.

## LIGHTING – EXTERIOR

### LIGHT & DESIGN ASSOCIATES

**Unit 0615, Bell House, 49 Greenwich High Road, London SE10 8JL**
**Tel 020 8469 4000**
**Fax 020 8469 4005**
**E-mail design@lightanddesign. co.uk**
Interior and exterior lighting specialists. Architectural lighting design consultants, established in 1990, specialising in the interior and exterior lighting of historic buildings including churches, museums and palaces. Lighting design awards have been gained for new lighting systems at St Luke's, Battersea, and Hinde Street Methodist Church, Marylebone. Recent commissions include concept proposals for exterior lighting at Buckingham Palace, the re-lighting of the interior to The Chapel Royal, St James's Palace, façade lighting to the Royal Exchange building, London and the re-lighting of the interior to St Bartholomew the Great, Smithfields.

## MASONRY CLEANING

### BURLEIGH STONE CLEANING & RESTORATION CO LTD

*See advertisement on page 19.*

### BURNABY CLEANING CO LTD

*See advertisement on page 19.*

## MEASURED SURVEYS

### SELWOOD AND DUNCAN

**Architectural Survey and Record**
**14 Lambridge, Bath BA1 6BJ**
**Tel/Fax 01225 421550**
**E-mail sd@surveyandrecord.co.uk**
**www.surveyandrecord.co.uk**
Specialists in recording ancient and historic buildings and churches. Accurate and comprehensive measured surveys, photographic recording, historical and documentary research. We use refined techniques of hand, instrument and electronic measurement, selected and integrated for an intelligent solution. Our surveys are an effective tool for projects and repairs, research and archiving.

## MORTARS & RENDERS

### BLEAKLOW INDUSTRIES

*See advertisement on page 32.*

### ST ASTIER NATURAL HYDRAULIC LIMES

*See advertisement on page 32.*

### THE TRADITIONAL LIME CO

*See advertisement on page 32.*

### TWYFORD LIME PRODUCTS

**1 Twyford Place, Tiverton, Devon EX16 6AP**
**Tel/Fax 01884 255407**
**E-mail arhunt@ 1twyford.fsnet.co.uk**
Mortars and renders. Manufacturer of lime putty, lime mortars, plasters and lime washes, skills for the repair and maintenance of ancient buildings, conservation and repair of cob buildings.

## ORGAN BUILDERS & RESTORERS

### HARRISON & HARRISON LTD

**St John's Road, Meadowfield, Durham DH7 8YH**
**Tel 0191 378 2222**
**Fax 0191 378 3388**
H & H are well known for large and distinguished instruments with electro-pneumatic action, and also for excellent organs with tracker-action. They are ready to work anywhere in the world. Even in a busy city church a Harrison organ will last many years before needing any significant work. Restoration of good organs is also a speciality.
*See also the Organ Builders and Restorers Map on page 47.*

## PAINT ANALYSIS

### LISA OESTREICHER

**Jubilee House, High Street, Tisbury, Wiltshire SP3 6HA**
**Tel 01747 871717**
**Fax 01747 871718**
A full range of analytical skills and techniques for the study of paint and interior finishes within historic buildings provided, including identification of pigments and media and archival research. Full reports prepared to provide a detailed insight into the historical development of interior and exterior decorative schemes, for documentation purposes, conservation or accurate restoration. Assistance in the design and implementation of programmes of conservation and decoration. Clients include the National Trust, English Heritage, architects, conservators and owners.

## PAINTS & FINISHES

### CLASSIDUR

**Blackfriar Paints, Blackfriar Road, Nailsea, Bristol BS48 4DJ**
**Tel 01275 854911**
**Fax 01275 858108**
*See advertisement on page 25.*

## PUBLICATIONS

### THE BUILDING CONSERVATION DIRECTORY

*See advertisement on page 48.*

### CHURCH BUILDING MAGAZINE

*See advertisement on page 48.*

### CONTEXT MAGAZINE

*See advertisement on page 48.*

### INSTITUTE OF FIELD ARCHAEOLOGISTS YEARBOOK

*See advertisement on page 51.*

### www.buildingconservation.com

The Building Conservation Directory's website **www.buildingconservation.com**, in association with Blackwell's on-line bookshop, offers a unique selection of book titles for historic building conservation, refurbishment and maintenance. Learn more about the history and protection of our architectural heritage by clicking in to www.buildingconservation.com. Select and order your books direct via the Internet using Blackwell's secure credit card purchasing system.
*See advertisement on inside back cover.*

## QUANTITY SURVEYORS

### BARE, LEANING & BARE

**2 Bath Street, Bath,
Somerset BA1 1SA**
**Tel 01225 461704**
**Fax 01225 447650**
**E-mail blbbath@btinternet.com**
**▶ Exeter Office Tel 01392 272245**
**Fax 01392 412089**
The partnership has extensive experience in the repair, alteration and extension of churches and church owned properties throughout the UK. The full range of QS services are offered together with costing of quinquennial reports, grant aid applications, planning supervision and VAT liability advice. Clients include many cathedrals and Anglican, Roman Catholic and non-conformist churches.

### SHAMBROOKS

**10 Clayfield Mews,
Newcomen Road, Tunbridge Wells,
Kent TN4 9PA**
**Tel 01892 540399**
**Fax 01892 540416**
Chartered quantity surveyors, building cost consultants and historic building consultants. Founded in 1942, Shambrooks have considerable experience of conservation and repair, and of installing services in historic buildings. Clients include English Heritage, The Church Commissioners, Royal Parks and Canterbury Cathedral. Services include grant applications, costing of listed building work and quinquennial maintenance, and Planning Supervision.

## ROOF DRAINAGE

### HARGREAVES FOUNDRY LTD

*See advertisement on page 27.*

### J & J W LONGBOTTOM LTD

**Bridge Foundry, Holmfirth,
Huddersfield HD7 1AW**
**Tel 01484 682141**
**Fax 01484 681513**
Long established foundry producing cast iron rainwater goods, gutters, pipes and fittings, air bricks and ornamental heads. Ex stock service on all standard items. Special requirements including curved gutters made promptly. Catalogue available on request.
*See entry in Heating section, page 56.*

### V & A TRADITIONAL LEAD CASTINGS LTD

*See advertisement on page 27.*

## ROOFING CONTRACTORS & MATERIALS

### AIRE VALLEY ROOFING

**22 Grange Park Drive, Bingley,
West Yorkshire BD16 1NR**
**Tel 01274 568878/560840**
**Fax 01274 560840**
Aire Valley Roofing has been specialising in all aspects of Yorkshire stone slate, Westmorland, Burlington and Welsh slate for 25 years. Most re-roofing works carried out in the Yorkshire region to a variety of buildings using Aire Valley's team of experienced slaters.

### CEL ARCHITECTURAL METAL ROOFING

*See advertisement on page 27.*

### KARL TERRY ROOFING CONTRACTORS LTD

**1 Stumbletree Cottages,
Ashford Road, Hamstreet,
Kent TN26 2EA**
**Tel/Fax 01233 732502**
**E-mail karl.terry@kentpegs.com**
**www.kentpegs.com**
Karl Terry Roofing Contractors Ltd specialises in old Kent peg tiling and roofing work to listed and period properties throughout South East England and nationwide. Fully experienced in tiling, slating and leadwork, directly employed craftsmen work to exacting standards. Clients include several oak-frame building and renovation specialists, and the company recently completed the roofing of an oak-framed building in Florence. Visit our website at www.kentpegs.com

### SALMON (PLUMBING) LIMITED

**Wentworth House, 24 Brox Road,
Ottershaw, Surrey KT16 0HL**
**Tel 01932 875050**
**Fax 01932 872972**
Traditional metal roofing in lead, copper, zinc and terne coated stained steel. Design and specification advice given, and new and refurbishment work undertaken for churches, schools and listed properties. Salmon Plumbing has large and small works divisions to suit the requirement. Established members of the Lead Contractors Association and Metal Roofing Contractors Association.

## SERVICES ENGINEERS

### THE CHAPMAN BATHURST PARTNERSHIP LTD

**32 St George's Place, Canterbury,
Kent CT1 1UT**
**Tel 01227 766172**
**Fax 01227 470122**
**E-mail cbp@
chapmanbathurst.co.uk**
For over 28 years we have enjoyed working on many projects, finding sympathetic solutions to the challenges of providing modern services for historic buildings. We advise the Diocese of Canterbury and are proud of our involvement with the restoration of Tonbridge Chapel and the £2.1 million extension to All Saint's Church, Crowborough, St Clement Danes and others.

### E C C CONSULTANTS

**The Hollies, The Green,
Upton, Norwich NR13 6BQ**
**Tel 01493 752232**
**Fax 01493 750965**
**E-mail ecc.con@which.net**
E C C Consultants has specialised in the design of heating and lighting systems for historic buildings for over 20 years. Recent projects include: Harwell Parish Church; Holy Trinity, Boston; Christ Church, Cockfosters; and the Roman Catholic Shrine at Walsingham. The firm is independent and impartial. Information and advice is available at all times and an informal visit to discuss your requirements is free.

## STEEPLEJACKS

### A C WALLBRIDGE & CO LTD

**Windsor Road, Salisbury,
Wiltshire SP2 7DX**
**Tel 01722 322750**
**Fax 01722 328593**
Wallbridge provides a comprehensive and professional steeplejack and lightning conductor service, carried out by the company's own staff of trade certified engineers who comply with current Health and Safety Regulations and British Standards. We are full members of the National Federation of Master Steeplejacks and Lightning Conductor Engineers.

### G & S STEEPLEJACKS LTD

**Tyne Depot, Stowey Road, Clutton,
Bristol BS39 5TG**
**Tel 01761 453516**
**Fax 01761 451557**
G & S Steeplejacks Ltd is a company very experienced in high-level repairs to historic and ecclesiastical buildings. Work includes masonry repairs to towers and spires, carving new stone, professionally undertaking general roof repairs in lead or copper, and installing, repairing and testing lightning conductors. The company also regularly re-paints or goldleafs clockfaces, and repairs and fits stained glass windows in all sorts of awkward places. G & S Steeplejacks Ltd also supplies and fits flagstaffs.

## PRODUCTS AND SERVICES

### W D REES STEEPLEJACKS – Lightning Protection Installation & Maintenance

**7 Mallard Close, St Mellons,**
**Cardiff CF3 0AJ**
**Tel 029 2079 6477**
**Fax 029 2030 8606**
**E-mail steeplejacks@buildwales.com**
**www.buildwales.com**

W D Rees Steeplejacks undertakes all types of high-level work including: church spires; stone repairs; new roofing and repairs; lightning conductor installation, testing and repairs; and repair and replacement of bells. All related work considered. Please contact Wayne Rees for further details or to discuss your project.

## STONE

### ANCASTER ARCHITECTURAL STONE LTD

*See advertisement on page 19.*

### ASHBY STONE MASONRY LIMITED

**Ashby Yard, 7 Grove Road,**
**Northfleet, Near Swanscombe,**
**Kent DA11 9AX**
**Tel 01474 327310**
**Fax 01474 364640**
**E-mail enquiries@ashbystone.co.uk**

Stonemasonry and restoration contractors. Ashby Stone Masonry Limited provide a full restoration service as principal contractor or sub-contractor from consultancy, survey and design, through procurement and production to site installation and repairs. All aspects of the conservation of buildings are encompassed to meet the project requirements to the highest standards.

### BURLEIGH STONE CLEANING & RESTORATION CO LTD

**The Old Stables, 56 Balliol Road,**
**Bootle, Merseyside L20 7EJ**
**Tel 0151 922 3366**
**Fax 0151 922 3377**
**E-mail info@burleighstone.freeserve.co.uk**
**www.burleighstone.co.uk**

*See advertisement on page 19.*

### CARTHY CONSERVATION LTD

**18 Alexandria Road,**
**London W13 0NR**
**Tel/Fax 020 8840 3294**
**E-mail deborahcarthy@btclick.com**

Carthy Conservation carries out high quality conservation and consultancy projects on stone, terracotta, plaster, mosaic and wood. Gilding, polychrome and monochrome surfaces are also specialities. Strong links established with analytical laboratories and scientists in the UK and abroad. Work is for private clients, architects and building consultants, main contractors, cathedrals, churches and government agencies.

### G A NEWTON (STONE)

**18 Nackington Road, Canterbury,**
**Kent CT1 3PN**
**Tel 01227 450364**
**Fax 01227 450364**

Sensitive, practical advice and repairs to historic stonework. An extensive range of practical skills and experience is combined with a creative, progressive approach to conservation and restoration work. Services include surveys, analysis and reports (specialising in monuments and tombs), masonry, carving, lettering and design. Recent clients include English Heritage and Churches Conservation Trust.

### McMARMILLOYD LTD

**The Old Farmyard, Brail Farm,**
**Great Bedwyn, Marlborough,**
**Wiltshire SN8 3LY**
**Tel 01672 870227**
**Fax 01672 870053**
**E-mail info@mcmarmilloyd.co.uk**
**www.mcmarmilloyd.co.uk**

Alabaster specialists and leading wholesalers of first quality stone, marble and granite etc. McMarmilloyd also specialises in statuary and in the sourcing and supply of rare, antique and ancient marbles and alabaster for refurbishment and restoration of important national monuments. Currently supplying material to The Ladbroke Grove Memorial, The Royal Society and the John Soane Mausoleum.

### MOTT GRAVES PROJECTS LTD

**Sampleoak Lane, Chilworth,**
**Guildford GU4 8QW**
**Tel 01483 453326**
**Fax 01483 453325**
**www.mottgraves.co.uk**

This company provides a comprehensive masonry service for private and professional clients. Specialising in the conservation and restoration of historic and listed buildings, they offer expertise in all aspects of stone masonry and project management. The directors personally organise and manage each commission, carefully combining proven restoration methods with a sensitive approach to new work. Previous works include the conservation and reconstruction of the Orangery and Coach House at Ingress Abbey, Dartford.

*See advertisements on pages 28 and 32.*

### NIMBUS CONSERVATION LTD

**Eastgate, Christchurch Street East,**
**Frome, Somerset BA11 1QD**
**Tel 01373 474646**
**Fax 01373 474648**
**E-mail enquiries@nimbusconservation.com**
**www.nimbusconservation.com**

Stone, marble and plaster conservators working on churches, cathedrals, monuments, tombs and wall paintings. They also carry out masonry, sensitive carving, specialist cleaning, analyses and detailed conservation reports in conjunction with Nimbus Consultancy. Recent contracts include Ardfert and Arundel Cathedrals, Bath Abbey and many churches in all areas of England, Ireland and Wales. Clients include The Churches Conservation Trust, Council for the Care of Churches, English Heritage, National Trust and PCCs.

### PAYE STONEWORK & RESTORATION LTD

**44-46 Borough Road, London SE1 0AJ**
**Tel 020 7928 4000**
**Fax 020 7928 4004**
**E-mail dm@payestone.co.uk**
**www.payestone.co.uk**

PAYE is a leader in its field, undertaking maintenance and repair of historic listed facades, regularly employed as principal contractor. Contracts include, Windsor Castle, Southwark Cathedral and the Tower of London. PAYE provides surveys, budgets and technical advice for clients establishing scopes of work and cost plans on future projects.

*See advertisement on page 16.*

### PRIEST RESTORATION

*See advertisement on page 16.*

### RUSSCOTT CONSERVATION & MASONRY

*See advertisement on page 16.*

### STONE FEDERATION GREAT BRITAIN

**Construction House,**
**56-64 Leonard Street,**
**London EC2A 4JX**
**Tel 020 7608 5094**
**Fax 020 7608 5081**
**www.stone-federationgb.org.uk**

The Stone Federation is the leading body in the Natural Stone industry. Its members operate quarries, supply and fix masonry and cladding, and provide cleaning, repair and restoration services. It promotes the use of natural stone and strives for the attainment of the highest standards, promoting training, health and safety, and developing technical standards. Please contact the Federation for more information.

### WELDON STONE ENTERPRISES LTD

**106 Kettering Road, Weldon,**
**Corby, Northants NN17 1UE**
**Tel 01536 261545**
**Fax 01536 262140**

Weldon Stone is a well-established company with experience in the restoration and conservation of churches, stately homes, and other listed buildings. They take pride in producing accurately sawn and worked masonry, fixed by teams of masons who understand the importance of careful handling and traditional techniques. Recent contracts include Stoneleigh Abbey, Warks (East Wing and Laundry House); St Mary Magdalene, Enfield, London; and St Botolph's, Ratcliffe-on-the-Wreake, Leicestershire.

### WELLS CATHEDRAL STONEMASONS LTD

**Brunel Stoneworks, Station Road, Cheddar, Somerset BS27 3AH**
**Tel 01934 743544**
**Fax 01934 744536**
**E-mail wcs@stone-mason.co.uk**
**www.stone-mason.co.uk**

Wells Cathedral Stonemasons are renowned for their high-quality masonry work on many of England's finest cathedrals and important ecclesiastical and historic properties, including sympathetic new building work. They provide a comprehensive range of masonry services covering surveying, design, carving, restoration, conservation and cleaning.

### WELLS MASONRY SERVICES LTD

**Ilsom Farm, Cirencester Road, Tetbury, Glos GL8 8RX**
**Tel 0166 650 4251**
**Fax 0166 650 2285**

A highly regarded company specialising in stone restoration to churches and listed buildings. Their fully equipped banker shop produces moulded and carved stone of the highest quality. The company also carry out stone/brick cleaning and offer stone selection advice and a full design service.

## STRUCTURAL ENGINEERS

### BLACKETT-ORD CONSULTING ENGINEERS

**33 Chapel Street, Appleby-in-Westmorland, Cumbria CA16 6QR**
**Tel/Fax 01768 352572**

The practice specialises in the repair of historic and traditional buildings in the north of England and southern Scotland. Recent church projects have involved bell frames and bell towers, masonry movement and decay, and timber and concrete roof repairs. Clients include Christ Church, Hartlepool; Liverpool Anglican Cathedral and Our Lady and St Hubert, Great Harwood.

### CAMERON TAYLOR BEDFORD
Consulting Structural Engineers

**Lorne Close, Regents Park, London NW8 7JJ**
**Tel 020 7262 7744**
**Fax 020 7724 0917**
**E-mail clive.richardson@cameron taylor.co.uk**
**▶ Birmingham, Bury St Edmunds, Chelmsford, Leeds, Norwich and Reading**

Specialists in the survey, repair and development of buildings, large and small. Expert advice for planning inquiries and litigation. Clients include the National Trust, the Church Commissioners, and the Crown Estate. CTB's dedicated conservation team is known to English Heritage, and led by Clive Richardson, Engineer to Westminster Abbey, and visiting lecturer in building conservation for the Architectural Association.

### THE MORTON PARTNERSHIP LTD
Structural and Civil Engineers

**The Old Cavalier, 89 Dunbridge Street, Bethnall Green, London E2 6JJ**
**Tel 020 7729 4459**
**Fax 020 7729 4458**
**▶ Arcadia House, 19 Market Place, Halesworth, Suffolk IP19 8BB**
**Tel 01986 875 651**
**Fax 01986 875 085**

Brian Morton founded this practice in 1966. It is now completely involved in minimum repair solutions to preserve historic buildings. Brian Morton is Engineer to Canterbury Cathedral. The practice has worked on some 300 churches and cathedrals. Preliminary advice and structural surveys are part of its work. All forms of conservation construction: brick, stone, timber and iron are part of the everyday business of the practice.

## STRUCTURAL REPAIRS

### HELIFIX LTD

**21 Warple Way, London W3 0RX**
**Tel 020 8735 5200**
**Fax 020 8735 5201**
**E-mail info@helifix.co.uk**
**www.helifix.co.uk**

Cost-effective and sympathetic structural repair and restoration techniques using the well-proven Helifix range of specially engineered, stress-free, stainless steel ties, fixings and reinforcement products. Suitable for use with all stone and masonry structures, Helifix concealed, non-disruptive repairs avoid taking down and rebuilding and are ideal for listed, historic and modern churches.

## SURVEYORS

### DIXON WEBB

*See advertisement on page 15.*

### HISTORIC BUILDING SURVEYS LTD (HoBS)

**Rose Ville, Main Street, Farrington Gurney, Bristol BS39 6UB**
**Tel 01761 451019 Vicky Yarham**
**E-mail jheward@netcomuk.co.uk**

Specialists in digital survey and analytical building recording, combining 30 years experience with the latest technology. We undertake detailed, high quality surveys of historic buildings and sites, watching briefs during restoration, conservation plans and documentary research. Clients include National Trust, Council for the Care of Churches and the Dean and Chapter of Ely Cathedral.

## TEXTILES

### FIONA HUTTON TEXTILE CONSERVATION

**Ivy House Farm, Wolvershill Road, Banwell, Somerset BS29 6LB**
**Tel 01934 822449**
**Fax 01934 822449**

We offer a full range of conservation treatments in our custom-designed studio, for antique or modern textiles damaged by accident or by poor storage or display conditions, including painted and printed textiles, woven tapestries, embroideries, upholstery and costume.

### WATTS & CO LTD

**7, Tufton Street, Westminster, London SW1P 3QE**
**Tel 020 7222 7169**
**Tax 020 7233 1130**
**E-mail enquiries@wattsandcompany.co.uk**
**www.wattsandcompany.co.uk**

Watts offers a famous collection of historic 19th and 20th century architect-designed textiles. Available in silk, wool and cotton, in authentic colours. These textiles are invaluable in the recreation of period interiors of churches and buildings of the past 200 years.

## TIMBER FRAME BUILDERS

### McCURDY & CO

**Manor Farm, Stanford Dingley, Reading, Berkshire RG7 6LS**
**Tel 0118 974 4866**
**Fax 0118 974 4375**
**E-mail info@mccurdyco.com**

With over 25 years experience, McCurdy & Co is one of the UK's leading specialists in the repair and conservation of historic timber frame buildings, together with the design and construction of historical and contemporary new timber framed buildings. For more information visit www.mccurdyco.com. *See also* The Heston Lych-gate *on page 33.*

## INDEX OF ADVERTISERS